IPv6
ネットワーク構築実習

前野 譲二・鈴田 伊知郎・小林 貴之　著

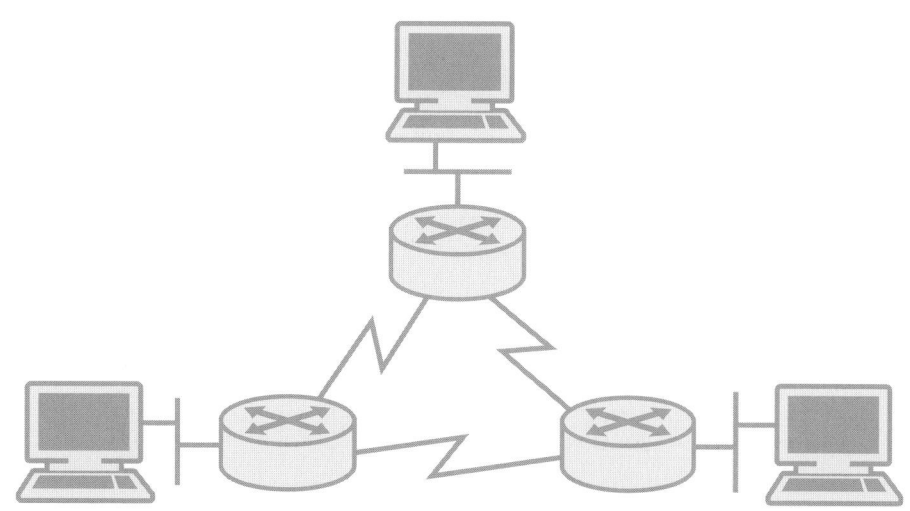

共立出版

本書によせて

　いまや社会インフラとなったインターネットですが，その日本での発展を見てきた私たちにとって大きな転換期となったのが 2011 年でした．それまで，ずっと使い続けてきた，かつ今でも使っている IPv4 のアドレスが枯渇したのです．昭和 40 年代（こういう書き方もすでにリアリティがない？）から言われている「石油」の枯渇や「土地」がそうであるように，有限なのに無くならない，と言われていたその枯渇が現実のものとなりました．とはいえすぐにインターネットの世界でなにか困っているわけではないのですが，ネットワークを学ぼうとする皆さんにとって IPv4 に変わる新しいプロトコルである IPv6 について学ぶ必要性はかなり高まっているといえます．

　すでにその技術は 2000 年頃から広く知られてきたわけですが，様々な経緯があって未だ普及しているとはいえない状況です．私たちはネットワーク技術そのものを研究対象にしているわけではありませんが，ネットワーク技術教育に携わる立場にあるとき，IPv6 についていかに学んでもらうかを常々考えてきました．現時点で普及していないからといって，今後必要となる技術は数年後に実務者となる皆さん（あるいはすでに，かもしれませんが）にはキチンと教える必要がある．そう思った時，技術解説書は書店にも数多く並んでいますが，実際に設定を通して学んでいく実習書がないことに私たちは気づいたのです．私たちが集った NPO・ILA(インターネット・ラーニングアカデミー)は学校教育でのＩＴ技術者の育成やそのための学校支援，また新しい学習環境の将来像を示すための研究，実証実験などを，自治体や企業，学校・教員と協力して行ってきました．特にネットワーク技術者養成教育としてあるメーカーの認定技術者資格習得を目標としたカリキュラム展開の中で，当初は IPv6 の技術理解を深め，そして伝えようと考えていたのですが，ここ数年は「IPv6 について実習を通して学べるものをつくろう」とのより具体的な思いで活動してきました．その意味で私たちの思いをこの本にまとめることができたのは ILA の助成や，メンバーや会員からのアドバイスなどがあったからこそといえます．

　本書はあくまで，教育現場で使われることを前提としましたので，機材も入手しやすい機器を選択しましたし，プロトコル自体についてはよりよい書籍があることですから簡単な解説にとどめています．できるだけ多くの方に，実際に機器の設定をすることを通して IPv6 についての理解が深まることを願っています．

　本書の各実習はルーターを使うものに関しては，各章の前半で異機種のルーターを相互に使う形式の実習を，後半でシスコルーターのみで行う形式の実習を用意してあります．特に前半の異機種でのルーターを相互に使う実習は，実際のインターネットにおいては対向の機器が異なるメーカーや機種であることはあたりまえのことですし，さらには同一の目的の設定をするにも，各メーカーの設計思想などが現れていて，非常に興味深いと思います．この点については私たちも企画段階から注目していた点です．各ルーターの OS がバージョンアップされる際に，変更を目の当たりにしたこともあり，まだ IPv6 の機能実装について揺れ動いていることを実感してきました．

　話がそれましたが，実習は異機種間構成，シスコ機種のみの構成，どちらもそれぞれ独立して進めていただいても問題ありません．ただ，本書を手にされる方は IPv4 においてそれなりの経験をつまれ

ていることが多いでしょうから，順番の前後は別にして，ぜひ異機種間構成を経験されることをお勧めします．単一メーカーのみの経験では得られない気づきがあるでしょう．

ただし，前述のとおりあくまでも教育機関で入手しやすい機器を選定したため，OSPFv3 の実習で 3 メーカーによる構成が取れませんでしたし，またシスコルーターの Ethernet インターフェースの数の関係から，別機種を用いていたりもします．この辺りも IPv6 の普及とともに時間が解決するかもしれませんが，現時点ではお許しいただければと思います．

私たちも前述のような課題を抱えていることもあり，今回の形が完成形とは考えておりません．この本を手にとっていただき，教育に利用いただく先生，そして実際に実習を行う皆様の存在が私たちの今後のモチベーションとなります．可能な範囲で忌憚のないご意見をお寄せいただければ幸いです．

最後になりますが，本書の出版にあたっては最終的には ILA の活動から離れた形になるなど紆余曲折もあり，計画よりはかなり遅れてしまったのですが ILA 幸友子さんの原稿整理をはじめ，共立出版株式会社編集制作部の吉村修司さん，DTP 作業を担当された祝竜平さんのご協力で何とか辿り着くことができました．ある程度のテキストの形ができた段階で次の目標を定められず，常日頃より毎回の作業が目的なのか，その後の深夜までの酒を飲みながらの議論が目的なのかわからないような私たちに出版という機会を与えてくださったことには本当に感謝しております．

2013 年 2 月

著者一同

本書の構成

第 1 章は IPv6 の概要について簡単にまとめてあります．実習は第 2 章からですが，すでにルーターの設定について基本的な知識がある方には不要です．実際の IPv6 ネットワーク構築は第 3 章から始まります．第 3 章の設定内容は他の実習における基本設定であり，他の実習開始時の起点となりますので設定内容を保存しておくことをお勧めします．第 4 章と第 5 章は静的経路の設定実習で，第 6 章からルーティングプロトコルの RIPng や OSPFv3 の実習を行います．また各ルーティングプロトコルの後半にはデフォルトルートをルーティングプロトコルで配布する実習が用意されています．実際のインターネット接続には ISP を経由しますが，ISP をデフォルトルートとして設定することが多いと思われますので，本実習を用意しました．第 10 章では IPv6 活用に関する実習として，IPv4 ネットワーク経由での IPv6 利用や IPv6 を利用したサーバー構築の実習を行います．最後の第 11 章は IPv6 セキュリティの実習としてアクセスコントロールリストと IPsec に関する実習を行います．

CONTENTS

Chapter 01　IPv6の基礎　001

- 01-1　はじめに …………………………………………………………… 002
- 01-2　IPv4アドレスとIPv6アドレスの相違点 ……………………… 003
- 01-3　IPv6アドレスの種類 …………………………………………… 004
- 01-4　IPv6アドレスの割り当て方法 ………………………………… 006
- 01-5　IPv6ルーティング ……………………………………………… 007
- 01-6　IPv6を利用したインターネットサービス …………………… 009
- 01-7　IPsec ……………………………………………………………… 009
- 01-8　IPv4経由でのIPv6トランスポート技術 ……………………… 010

Chapter 02　ルーターの基本操作　011

- 02-1　本書で用いるルーターとその基本的操作 …………………… 012

Chapter 03　IPv6の基本的な設定　019

- 03-1　マルチベンダー機器による実習 ……………………………… 020
- 03-2　シスコ機器による実習 ………………………………………… 031

Chapter 04　スタティックルートの設定　041

- 04-1　マルチベンダー機器による実習 ……………………………… 042
- 04-2　シスコ機器による実習 ………………………………………… 046

Chapter 05　デフォルトルートの設定　051

- 05-1　マルチベンダー機器による実習 ……………………………… 052
- 05-2　シスコ機器による実習 ………………………………………… 056

CONTENTS

Chapter 06 RIPng の設定　061
06-1　マルチベンダー機器による実習 …………………………………… 062
06-2　シスコ機器による実習 ……………………………………………… 067

Chapter 07 RIPng のデフォルトルートの伝搬　073
07-1　マルチベンダー機器による実習 …………………………………… 074
07-2　シスコ機器による実習 ……………………………………………… 081

Chapter 08 OSPFv3 の設定　091
08-1　マルチベンダー機器による OSPFv3 基本設定 ………………… 092
08-2　シスコ機器による OSPFv3 基本設定 …………………………… 106

Chapter 09 OSPFv3 のデフォルトルートの伝搬　117
09-1　マルチベンダー機器による OSPFv3 のデフォルトルートの伝搬 … 118
09-2　シスコ機器による OSPFv3 のデフォルトルートの伝搬 ……… 135

Chapter 10 IPv6 活用　149
10-1　IPv6 over IPv4 トンネリングの設定 …………………………… 150
10-2　IPv6 実習サーバーの構築 ………………………………………… 159

Chapter 11 IPv6 セキュリティ　167
11-1　IPv6 ACL ……………………………………………………………… 168
11-2　IPv6 での IPsec ……………………………………………………… 172

Chapter 01

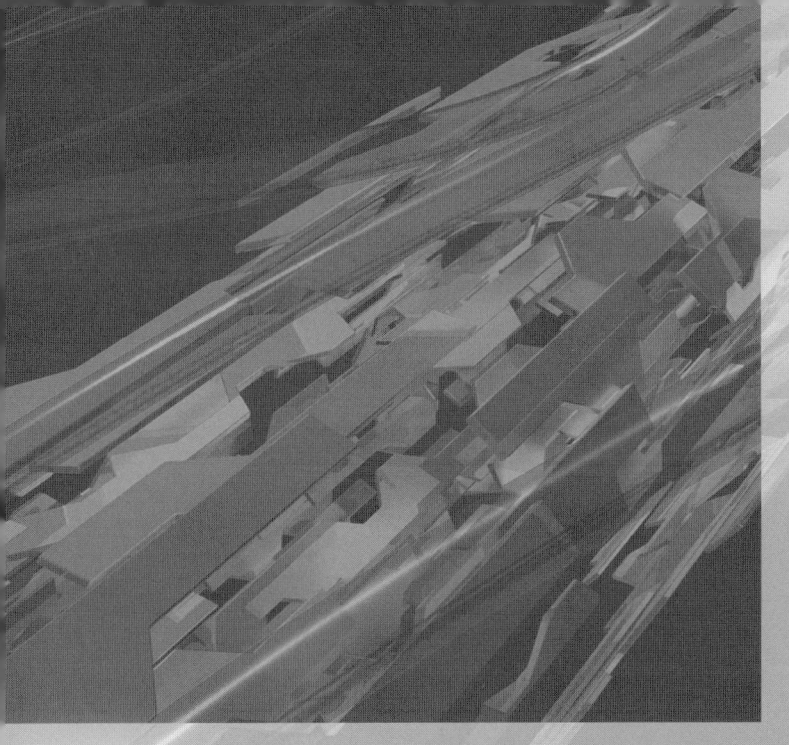

IPv6の基礎

01-1 はじめに

01-2 IPv4アドレスと
　　　IPv6アドレスの相違点

01-3 IPv6アドレスの種類

01-4 IPv6アドレスの
　　　割り当て方法

01-5 IPv6ルーティング

01-6 IPv6を利用した
　　　インターネットサービス

01-7 IPsec

01-8 IPv4経由での
　　　IPv6トランスポート技術

Chapter 01 IPv6の基礎

01-1 はじめに

　コンピュータネットワークの世界で利用されている通信規約（プロトコル）には，国際的な標準化委員会 ISO による OSI の 7 階層モデルと，業界標準である TCP/IP の 4 階層モデルが知られています。これら 2 つの通信規約モデルにおいて，OSI 参照モデルの 3 層目と TCP/IP モデルの 2 層目にそれぞれ割り当てられているのがネットワーク層です。そして，IP（Internet Protocol）インターネットプロトコルは，ネットワーク層で利用される通信規約です。

　IP は，インターネット上でデータをパケットとして伝送する際に使用され，特に IP アドレスはノードの識別に利用される論理アドレスとして有名です。これまでバージョン 4 の 32 ビット長の IPv4 アドレスが利用されてきましたが，アドレス枯渇が想定されたため 128 ビット長の IPv6 アドレスが策定され，1999 年から実際に割り当てが開始されました。新聞などで既報ですが，新規割り当ての IPv4 アドレスがすべて払い出されたため，今後 IPv6 アドレスが広く利用されることになります。2012 年までの IPv6 アドレスに関して大きな節目となった年とその内容は表 1-1 に示すとおりです。

表 1-1　IPv6に関するできごと

年月	内容
1995年 1月	SIPPをベースにアドレスを128bit化 (RFC1752) IPngからIPv6に正式に改名
1995年 12月	IPv6アドレス体系決定 (RFC1884，その後1998年7月にRFC2373に改訂)
1996年 6月	日本で最初にIPv6パケットが専用線内 (Wide-6bone) で通信される (東大と奈良先端大間)
1998年末	IPv6関係RFCが大幅に改定される
1999年 8月	IPv6アドレス割り当て開始
2002年 8月	SINETがIPv6トンネリングサービス開始
2003年 9月	IPv6 Readyロゴ認定プログラム開始
2007年 12月	SINETがIPv6ネイティブ接続，デュアル接続サービス開始
2008年 2月	DNSの6つのルートゾーンにAAAAレコードが登録され，IPv6トランスポートによるルートDNSサーバーへの問い合わせが開始される
2011年 2月	IANAのIPv4在庫枯渇
2011年 4月	APNIC，JPNICのIPv4在庫枯渇
2011年 6月	World IPv6 Day
2012年 6月	World IPv6 Launch

01-2 IPv4アドレスとIPv6アドレスの相違点

IPv6アドレスはIPv4アドレスを拡張したものですが，表1-2にあるようにいくつか異なる点があります。

表1-2　IPv4とIPv6の相違

	IPv4	IPv6
アドレス長	32ビット長	128ビット長
アドレスの種類	ループバック ホスト	ループバック ホスト リンクローカル サイトローカル
アドレスタイプ	ユニキャスト マルチキャスト ブロードキャスト	ユニキャスト マルチキャスト エニーキャスト
アドレス割り当て方法	静的(手動) DHCP	静的(手動) DHCP RA
パケットヘッダ長	可変	固定

　IPv6ではネットワークに接続する機器に割り当てることができるアドレスの数を多くするため，ビット長を4倍にし，これにより接続できる機器数を大幅に増やすことができます。しかし，ビット長を128ビット，すなわち2進法で128桁にもなるアドレスの表記はIPv4アドレスとは異なります。IPv4アドレスでは32ビットを8ビットずつ4つのブロックに分割して，10進法で表記するのが一般的です。8ビットで表すことができる数値は10進法で0～255となります。IPv6も同様に8ビットずつに分割すると16ブロックに分割することになり，もっと少ないブロックにするには16ビットずつで8ブロックになります。16ビットで表すことができる数値は10進法で0～65535となりますが，16進法を用いれば0～ffffとなります。このため現在IPv6アドレスは16進法8ブロックで表すことが一般的です。

　アドレスの先頭からネットワークを識別するネットワークアドレスが記載され，後半にノードを識別するノードアドレスが記載されます。このとき，ネットワークアドレスのビット数（プレフィックス）を併せて表記します。IPv6のアドレス表記は以前RFC4291で仕様化されていましたが，現時点ではRFC5952で規定されています。この中で推奨されている表記方法は以下のとおりです。

- 16ビットフィールドの先頭の0は省略してください。0000の場合は0としてください。
- :: を利用して可能な限り0を省略してください。
- 16ビットフィールドの0フィールドが1つだけの場合，:: を使用して省略しないでください。
- :: を使用して省略可能なフィールドが複数ある場合，最も多くの16ビット0フィールドが省略できるフィールドを省略してください。また省略できるフィールド数が同数の場合は前方を省略してください。例えば2001：db8：0：cafe：0：0：1の場合，2001：db8：：cafe：0：0：1または2001：db8：0：cafe：：1と0を省略できますが，後半の2001：db8：0：cafe：：1の方が0を多く省略できるので推奨されます。
- 16進法のa～fは小文字を使用してください。

なお，アドレスにポート番号を併記する場合は以下の6通りが可能ですが，使用するソフトウェアによってはポート番号と解釈されない場合があります。

```
[2001:db8::1]:80
2001:db8::1:80
2001:db8::1.80
2001:db8::1 port 80
2001:db8::1p80
2001:db8::1#80
```

01-3 IPv6アドレスの種類

IPv6アドレスは，特定の1つのノード相手と通信するためのユニキャストアドレスと複数のノード相手と通信するためのマルチキャストアドレス，さらにエニーキャストアドレスに分けられます。

ユニキャストアドレスはさらにIPv4アドレスのグローバルアドレスに相当するグローバルユニキャストアドレス，IPv4アドレスのプライベートアドレスに相当するユニークローカルユニキャストアドレス，そしてIPv6独自のリンクローカルユニキャストアドレスがあります。

■ グローバルユニキャストアドレス

インターネット上で一意のアドレスです。先頭の3ビットは必ず001（2進法）で，次の45ビットがネットワークプロバイダから割り当てられ，計48ビットがグローバルルーティングプレフィックスと呼ばれます。次の16ビットはサブネットIDと呼ばれ，IPv6アドレスを割り当てられた各組織内で管理します。残りの64ビットはインターフェースIDと呼ばれ，ノードに割り当てます。

■ ユニークローカルユニキャストアドレス

組織内で自由に割り当てられるプライベートなアドレスです。このアドレスでインターネットと通信することはできません。したがって，実験的なネットワークや外部と通信を行わないネットワークで用います。アドレス最初の7ビットは1111110（2進法）で，8ビット目は0をIETFが予約済みのため原則1となります。次の40ビットはグローバルIDと呼ばれ推奨された乱数を用いた計算式により求められた数値とします。次の16ビットはサブネットIDと呼ばれ，各組織内で管理します。残りの64ビットはインターフェースIDと呼ばれ，ノードに割り当てます。

■ リンクローカルユニキャストアドレス

隣接ノードとの通信のみに利用され，ルーターを越えて通信することができないアドレスです。当然インターネットと通信することもできません。アドレスの最初の10ビットは1111111010（2進法）で，次の54ビットは未使用ですべて0となります。残りの64ビットはインターフェースIDと呼ばれ，ノードに割り当てます。一般にOSなどが自動的に割り当てていますが，手動で設定することも可能です。

■ マルチキャストアドレスとエニーキャストアドレス

マルチキャストとはTVやラジオなどの放送と同じように1対多数の通信形式です。マルチキャストアドレスはff00::/8のプレフィックスをもち，次の8ビットの上位4ビットが0000の場合IANAが

割り当てた恒久的なマルチキャストアドレスで，0001の場合は組織内で割り当てた一時的なマルチキャストアドレスとなります。下位4ビットはマルチキャストを送信する範囲で現在6種類が下記のように指定されています。

表1-3　マルチキャストアドレス

マルチキャストアドレス	スコープ範囲
ff01::/16	インターフェースローカルスコープ
ff02::/16	リンクローカルスコープ
ff04::/16	管理ローカルスコープ
ff05::/16	サイトローカルスコープ
ff08::/16	組織ローカルスコープ
ff0e::/16	グローバルスコープ

IPv6では，IPv4に存在するすべてのノードに対しての通信であるブロードキャストが無くなっています。またルーティングプロトコルで利用されるマルチキャストアドレスも表1-4のようにIPv4から変更されています。

表1-4　マルチキャストグループ

マルチキャストグループ	IPv4	IPv6
全ノード		ff01::1
全ルーター		ff01::2
同一リンク上全ノード	224.0.0.1/24	ff02::1
同一リンク上全ルーター	224.0.0.2/24	ff02::2
同一リンク上全OSPFルーター	224.0.0.5/24	ff02::5
同一リンク上全OSP　DR	224.0.0.6/24	ff02::6
RIP	224.0.0.9/24	ff02::9

IPv6アドレスは1つのインターフェースに複数かつ，他のノードと同じアドレスを割り当てることが可能です。このため同一のアドレスが複数のインターフェース上に存在する場合が生じます。このようなアドレスをエニーキャストアドレスと呼びます。エニーキャストアドレスに対して通信を行う場合，最もネットワーク的に近いノードと1対1で通信することになります。また，事前に割り当てられているエニーキャストアドレスとしてモバイルIPv6ホームエージェント(RFC2556)などが存在します。

01-4 IPv6アドレスの割り当て方法

　IPv6アドレスの機器への割り当て方法は，IPv4と同様に，手動割り当てと自動割り当てがあります。自動割り当てにはDHCPサーバーを利用するステートフル割り当てと，DHCPサーバーを利用しないステートレス割り当て，および，ステートフルとステートレスを併用する割り当ての3つの方法があります。

■ステートレス割り当て

　DHCPを利用しないステートレスアドレス設定は，128ビットのIPv6アドレスの前半64ビットをルーターから得られるプレフィックスとし，後半64ビットはMACアドレスをベースに生成するものです。48ビットのMACアドレスであるEUI-48をIEEEが定めたEUI-64に変換し，インターフェースIDとして使用します。48ビットのMACアドレスの前半24ビットの製造会社を示すカンパニーIDと後半24ビットの間にff:feを挿入します。さらに先頭から7ビット目を反転させます。これは先頭から7ビット目が示す内容が48ビットと64ビットで逆になっているためです。ただし，この方法では，アドレスからMACアドレスが逆算でき，ノードが特定されるなど，プライバシーの問題が生じるおそれがあります。このため，ランダムに割り当てる方法も提案されています。生成されたインターフェースIDを利用してリンクローカルアドレスを生成し，同一セグメント上に同一アドレスが存在しないことを確認後，RSを送信しルーターなどからアドレスプレフィックス情報をRAとして入手して，最終的ユニキャストアドレスを完成させます。ただし，この方法では，DNSサーバーの情報を得ることができないため，手動でDNSサーバーのアドレスを設定する必要があります。

■ステートフル割り当て

　DHCPを利用してアドレスを割り当てます。この方法ではDNSサーバー情報も自動的に設定できますが，デフォルトゲートウェイは手動で設定する必要があります。

■併用割り当て

　ルーターからのRA情報にステートレスアドレスプレフィックスが含まれていて，かつ，ホストがステートフルなアドレス構成プロトコルを使用している場合は，両方の情報が併用された割り当てとなります。

01-5 IPv6ルーティング

IPv6でのルーティングは，IPv4でのルーティングと同様に，静的ルーティングとルーティングプロトコルを利用した動的ルーティングが可能です。IPv6ネットワークのルーティングプロトコルは，IPv4ネットワークのルーティングプロトコルを拡張したものが一般的です。表1-5にIPv4とIPv6ルーティングプロトコルを比較しました。

表1-5　IPv4とIPv6のルーティング方式比較

ルーティング方式	IPv4	IPv6
静的	利用可	利用可
デフォルトルート	利用可	利用可
RIP	v1/v2	ng
OSPF	v2	v3
EIGRP	EIGRP	EIGRPv6
ISIS	ISIS	ISISv6
BGP	BGP4	BGP4+

■RIPng

RIPngは，RFC2080で定義されたIPv6によるRIPの実装です。RIPv1/v2の比較を表1-6に示します。表からわかるように多くの特徴が，RIPngにもそのまま引き継がれています。

表1-6　RIPの比較

	RIPng	RIPv1	RIPv2
最大ホップ数	15	15	15
ルーティング情報の宛先アドレス	リンクローカルマルチキャスト (ff02::9)	ブロードキャスト	ブロードキャストまたはマルチキャスト (224.0.0.9)
宛先ポート	UDP 521	UDP 520	UDP 520
経路アドバタイズ周期	30秒	30秒	30秒
エージングタイマー	180秒	180秒	180秒
ホールドダウンタイマー	120秒	120秒	120秒

ルーティングテーブルは30秒に1回アドバタイズされますが，大規模なブロードキャストネットワークではコリジョンを避けるために，次のいずれかを実装することが求められています。

　▲システムの負荷に影響されないクロックによって30秒ごとのアップデートを行う
　▲30秒のタイマーを，最大で±15秒のランダムな秒数を加減してアップデートを行う

RIPv1/v2同様，180秒間アップデートがない場合，そのルートは失われたものとみなされ，そのルートは削除されます。またそのルートのメトリック（ルーティングにあたって最適な経路を得る基準）が無限カウントを示す16に設定されている場合にも削除されます。

■OSPFv3

　OSPFv3は，IPv6で動作するリンクステート型ルーティングプロトコルです。OSPF for IPv6（OSPF v3）は RFC2740 で，IPv4 用の OSPF Version 2 は RFC 2328 でそれぞれ定義されています。

　リンク上で動作するのでインターフェース上に設定する必要があり，v2 と v3 の両方ともルーターID（32 ビット長）が必要となります。また OSPF では必ずバックボーンエリアを作成する必要があります。

　OSPFv3 で利用するマルチキャストは，リンクローカル上の全 SPF ルーター宛の ff02::5 とリンクローカル上の DR ルーター宛の ff02::6 です。ルーティングアップデートはリンクローカルアドレスが使用されます。また認証が必要な場合は v3 では，ルーティングプロトコル内で行うのではなく，IPsec を用いる必要があります。

■ISISv6

　ISIS は IPv6 で動作するリンクステート型ルーティングプロトコルです。ISIS は OSI モデルで利用されます。欧州と異なり OSI モデルの利用が少ない日本国内での利用はほとんどないと思われます。ISIS はバックボーンエリアを作成する必要はありませんが，NPSA アドレスが必要となります。

■BGP4+

　マルチプロトコル BGP (BGP4+) は BGP のマルチプロトコルバージョンで，RFC2283 で規定されています。OSPF と同様に 32 ビットのルーター ID が必要となります。

01-6 IPv6を利用したインターネットサービス

IPv4で利用されているWebやDNSなどのインターネットサービスは現在IPv6に対応しています。

■ DNSサービス

IPv6アドレスとFQDNの対応表をもつDNSサーバーのIPv6化は，IPv6に対応した正引きアドレス情報と逆引きアドレス情報をもつ必要があります。また，1つのDNSサーバーでIPv4とIPv6の両方に対応することもできます。以前はIPv4に関してのみサービスを提供するDNSサーバーがありましたが，現在ではほとんどIPv6にも対応しています。

■ Webサービス

ApacheやIISなどのWebサーバーもIPv6に対応しており，設定を行えばIPv6サービスを提供することが可能です。また，IPv4でアクセスしてきた場合とIPv6でアクセスしてきたときに異なる情報を表示させることも可能です。

01-7 IPsec

IPsecはパケットを暗号化する技術です。IPsecを利用することにより，データの機密性と通信先の認証が可能となります。IPsecにはデータ部分のみ暗号化するトランスポートモードと，パケットヘッダを含む全体を暗号化するトンネルモードがあります。IPsecを構成するパラメータには，AH (Authentication Header)，ESP (Encapsulated Security Payload) や IKE (Internet Key Exchange protocol) などがあります。第11章の実習でこれらのパラメータを設定します。

01-8 IPv4経由でのIPv6トランスポート技術

ネットワークの経路上にIPv6に対応せずIPv4のみに対応しているネットワークがある場合，そのネットワーク内にトンネルを作成しIPv4ネットワーク内にIPv6パケットを通して，通信先のIPv6ノードに接続する方法が利用されます。トンネル技術には，手動でトンネルを構成する方法と，表1-7にあるように自動的にトンネルを構成する方法があります。

表1-7　自動トンネルの種類

自動トンネル方法	特徴
6to4	IPv4グローバルアドレス必要，NAT利用時動作保証なし
Teredo	IPv4グローバルアドレス不要，NAT利用時もほとんど動作可能
ISATAP	IPv4グローバルアドレス不要

■6to4 (RFC3056)

IPv6アドレス中にグローバルのIPv4アドレスを埋め込んで利用します。このためプライベートアドレスしかもたないホストでは，この方法を利用することはできません。また，NAT機器が存在する場合，動作保証はされません。この方法はホストとホスト間またはホストとルーター間で利用され，サイトアドレスに2002::/16のプレフィックスを使用する必要があります。

■Teredo (RFC4380)

NAT配下のホストにも対応できる技術で，ホストとホスト間で利用されます。ただし，すべてのNAT技術に対応できるわけではなく，シンメトリックNATには対応しません。Teredoリレールーターを経由してIPv6ネットワークと通信を行います。また，プレフィックスとしては2001:0000::/32を使用します。

■ISATAP (RFC5214)

ISATAPは6to4と異なりIPv4プライベートアドレスしかもたないホストも利用することができます。必要な機器はIPv4/IPv6デュアルスタック運用可能なルーターです。ルーターがカプセル化されたIPv4からIPv6パケットを取り出して，IPv6ネットワークへ転送するものです。

なお，10-1節では手動でトンネルを構成します。

Chapter 02

ルーターの基本操作

02-1 本書で用いるルーターと
　　 その基本的操作

Chapter 02 ルーターの基本操作

02-1 本書で用いるルーターとその基本的操作

　この章では本書を通じて利用する，3機種のルーターについて基本的な操作を学びます。本章の目的はあくまで今後の実習に必要な最低限の操作を確認することです。初めてルーターを触れる方は少ないと思いますし，多くの方はシスコ製ルーターに慣れていらっしゃるのではないでしょうか。本書で用いるルーターは以下の機器を前提としています。

　　　Cisco1841（12.4(24)T3/Advanced IP Services）
　　　YAMAHA RTX1200（Rev.10.01.22）
　　　ALAXALA AX620R-2025（NEC IX2025）（ver8.3.46）

　選定にあたっては，教育機関で購入しやすい価格帯の機種を選定したつもりです。ただし，シスコについては章によってポート構成の関係でCisco1812Jを用いていることもあります。これはイーサネットインターフェースの実装ポート数によるものです。また，RTX1200についてはOSPFv3の実装がなされていないため，当該章では用いることができませんでした。AX620RについてはAX620R-2015でも実習可能ですし，そもそもAX620RはNEC IXシリーズのOEMであることからIX2025, IX2015でも同様に実習可能です。OSのバージョンについては動作確認したものを記載しました。このバージョンでないと実習できないということではありませんが，機種によってはバージョンにより細かいセマンティクスなどの差分が見受けられますので，実習を進める中で本書どおりに入力できないことが生じる可能性もあります。そこは各ルーターのヘルプ機能などを有効に活用いただきたいと思います。

　それではさっそく，準備に進みましょう。

STEP1　PCの設定

　PC側のシリアルポートについてはPCにオンボードで実装されることも少なくなったためUSB接続の変換器を用いることが多いと思われますが，詳細についてはここでは省略します。シリアルポートが用意されていてターミナルエミュレーターソフトがインストールされていることを前提に進めます。

　3機種ともターミナルエミュレーターソフトの設定は共通です。

表2-1　シリアルポート設定内容

ボー・レート	9600
データ	8bit
パリティ	none
ストップビット	1bit
フロー制御	none

　よく使われるターミナルエミュレーターソフトのTera Termでは図2-1のように［設定］-［シリアルポート］を設定します。（下の例ではシリアルポートがCOM5の場合）

02-1 本書で用いるルーターとその基本的操作

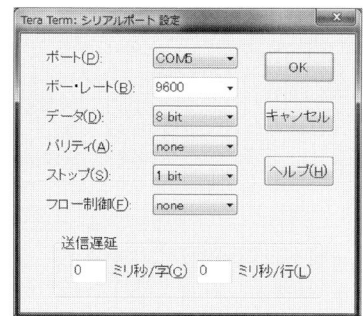

図2-1　Tera Termの設定

STEP2　ルーターとの接続

■ Cisco1841：

ケーブルは機器に添付のケーブルを用います。図2-2のケーブルが添付されているはずです。

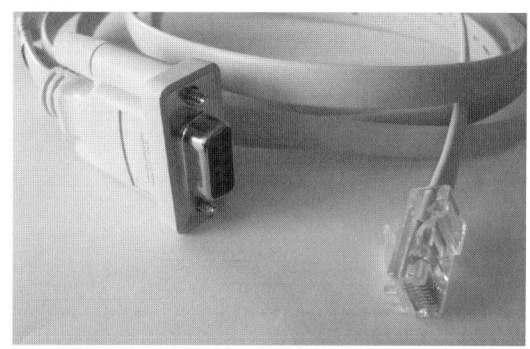

図2-2　シスコ用コンソールケーブル

ケーブルのRJ45側をCisco1841背面のコンソールポートに接続します。

■ RTX1200：

ケーブルは機器に添付されているものを用います（図2-3）。

図2-3　RTX1200用コンソールケーブル

ただし，このケーブルはシリアル (RS-232C：D-sub9pin) のクロスケーブルなので電器店などでも入手可能です。

接続に方向性はありませんので，片方を前面のコンソールポートに接続します。

■ AX620R:
ケーブルは機器に添付されているものを用います。図2-4のケーブルが添付されているはずです。

図2-4　AX620R用コンソールケーブル

両端の形態はシスコ用コンソールケーブルと同じですが，厳密な意味での互換性はありません。RJ45側を前面のコンソールポートに接続します。AX620R-2015（IX2015）を用いている場合は，ポートがプレートで隠されていますので，プレートを取り外してアクセスする必要があります。

STEP3　電源投入からコマンドプロンプトまで

ここでは工場出荷状態を前提として，設定可能となる状態のコマンドプロンプトが出るまでを実習します。すでに何らかの設定が入っている場合はSTEP4に進んでください。

■ Cisco1841:
起動時にはOSのメモリーへの展開などが行われ，その後Setupモードに入るか否かを尋ねられます。

```
System Bootstrap, Version 12.3(8r)YH12, RELEASE SOFTWARE (fc1)
Technical Support: http://www.cisco.com/techsupport
Copyright (c) 2007 by cisco Systems, Inc.
C1800 platform with 262144 Kbytes of main memory with parity disabled

     ( 中略 )

          --- System Configuration Dialog ---

Would you like to enter the initial configuration dialog? [yes/no]:
```

ここで **no** を入力し，再度 RETURN キーもしくは Enter キーを押下することを求められますので，それに従います。

```
Press RETURN to get started!
```

するとルーターに関する情報が出力されたのち，コマンドプロンプトが表示されて準備完了です。

```
Router>
```

■ RTX1200:
電源を投入すると，しばらくメッセージが表示されますが，メモリーの量とインターフェースの数が表示されます。

```
RTX1200 BootROM Ver.1.01
  Copyright (c) 2009 Yamaha Corporation

Press 'Enter' or 'Return' to select a firmware and a configuration.
Default settings :   exec0 and config0
```

```
(中略)
  Copyright (c) 1988-1992 Carnegie Mellon University All Rights Reserved.
00:a0:de:65:af:d1, 00:a0:de:65:af:d2, 00:a0:de:65:af:d3
Memory 128Mbytes, 3LAN, 1BRI
```

ここでRETURNキーを押下するとパスワードの入力が求められます。初期設定ではパスワードは設定されていませんので，さらにそのままRETURNキーを押下するとメッセージを表示したのちコマンドプロンプトが表示されます。

```
Password:

RTX1200 Rev.10.01.32 (Wed Jul 13 17:42:17 2011)
(中略)
00:a0:de:65:af:d1, 00:a0:de:65:af:d2, 00:a0:de:65:af:d3
Memory 128Mbytes, 3LAN, 1BRI
>
```

■ **AX620R:**

電源投入後，起動が完了するとオペレーションモードでかつアドミニストレーター権限のコマンドプロンプトが表示されます。

```
NEC Bootstrap Software
Copyright (c) 2001-2010 NEC Infrontia All Rights Reserved.

%BOOT-INFO: Trying flash load, exec-image [ix2025-ms-8.4.23.ldc].
Loading:######################################################## [OK]

(中略)

Copyright (c) 2001-2010 NEC Infrontia All Rights Reserved.
Copyright (c) 1985-1998 OpenROUTE Networks, Inc.
Copyright (c) 1984-1987, 1989 J. Noel Chiappa.
Router#
```

STEP4 モードとコマンドプロンプト

■ **Cisco1841:**

起動時のコマンドプロンプトの状態はRouter>となっているはずです。(Routerの部分はホスト名が設定してあれば，その文字列になります。)このコマンドプロンプトの最後が>となっている状態が**ユーザーモード**です。この状態でできることは限られていますのでイネーブルモードに移行します。

```
Router>enable
Router#
```

このコマンドプロンプトが#となっていることで現在イネーブルモードになっていることがわかります。さらに各種設定を行うためにグローバルコンフィグレーションモードに移行する必要があります。
そのためには **conf t** と入力します。

```
Router#conf t
Enter configuration commands, one per line.  End with CNTL/Z.
Router(config)#
```

コマンドプロンプトの最後が(config)#となっていることでグローバルコンフィグレーションモードになっていることがわかります。

イネーブルモードに戻るには **exit** と入力します。

```
Router(config)#exit
Router#
```

■ RTX1200：

RTX においてはモードという概念ではなく，アクセスレベルという考え方になります。

起動時はアクセスレベルが一般ユーザーになります。これはコマンドプロンプトが > となっていることからもわかります。ここで **administrator** と入力すると管理ユーザーのレベルとなり，コマンドプロンプトが変化します。パスワードは設定していなければ RETURN キーを押下します。

```
>administrator
Password:
#
```

この状態で RTX はルーターからのメッセージに日本語を表示することも可能ですが，環境により手間がかかることもあるので，ほかの機器と揃える意味でメッセージを英語表示にします。

```
#console character ascii
```

それから **exit** を入力することで一般ユーザーに戻ることができます。

```
#exit
>
```

■ AX620R：

起動時はオペレーションモードですのでコンフィグモードに移るには **enable-config** と入力します。あるいは **config** のみでもさらには **conf t** と入力しても移行可能です。オペレーションモードに戻るには **exit** と入力します。

```
Router#enable-config
Enter configuration commands, one per line. End with CNTL/Z.
Router(config)#exit
Router#conf t
Enter configuration commands, one per line. End with CNTL/Z.
Router(config)#exit
Router#
```

STEP5　設定の確認

今後，実習中に設定を確認する必要が生じますので，確認方法を次に示します。

■ Cisco1841：

イネーブルモードにて，**show running-config** コマンドで実行中の設定を確認できます。

```
Router#show running-config
Building configuration...

Current configuration : 1014 bytes
!
version 12.4
service timestamps debug datetime msec
(以下略)
```

また，保存されている設定は **show startup-config** コマンドで確認できます。

```
Router#show startup-config
startup-config is not present
```

上の例は，設定が保存されていない場合に表示されます。

■RTX1200:

管理ユーザーのアクセスレベルで **show configuration** コマンドで確認します。

```
#show configuration
# RTX1200 Rev.10.01.29 (Wed Feb  9 17:21:33 2011)
# MAC Address : 00:a0:de:37:5e:fd, 00:a0:de:37:5e:fe, 00:a0:de:37:5e:ff
# Memory 128Mbytes, 3LAN, 1BRI
# main:  RTX1200 ver=b0 serial=D26005478 MAC-Address=00:a0:de:37:5e:fd
  MAC-Address=00:a0:de:37:5e:fe MAC-Address=00:a0:de:37:5e:ff
# Reporting Date: Mar 24 11:47:15 2012
console character ascii
ip lan1 address 192.168.100.1/24
dhcp service server
dhcp server rfc2131 compliant except remain-silent
dhcp
```

また，保存されている設定の一覧を **show config list** コマンドで確認できます。

```
#show config list
No.  Date       Time      Size   Sects    Comment
-----  ----------  --------  -------  -------  ------------------------------------
*  0   2012/03/24 11:51:14    261  124/124
   0.1 2012/03/24 11:50:11    261  125/125
   0.2 2012/03/24 11:49:20    229  126/126
-----  ----------  --------  -------  -------  ------------------------------------
#
```

この内，起動時に読み込まれる設定は 0 なので，**show config** コマンドで確認できます。この際ログインパスワードと管理ユーザーのパスワードが必要になります。

```
#show config 0
Input passwords of CONFIG0
Login Password:
Administrator Password:
#10 1 29 0
ip lan1 address 192.168.100.1/24
pp disable all
no tunnel enable all
dhcp service server
dhcp server rfc2131 compliant except remain-silent
dhcp scope 1 192.168.100.2-192.168.100.191/24
#
```

■AX620R:

コンフィグモードにて，**show running-config** コマンドで確認できます。

```
Router(config)#show running-config
Current configuration : 818 bytes

! NEC Portable Internetwork Core Operating System Software
! IX Series IX2010 (magellan-sec) Software, Version 8.3.46, RELEASE SOFTWARE
! Compiled Feb 12-Fri-2010 10:44:00 JST #1
! Current time Mar 24-Sat-2012 10:07:55 JST
(以下略)
```

保存されている設定は **show startup-config** コマンドで確認できます。

```
Router(config)#show startup-config
% Non-volatile configuration memory is not present
```

上の例は，設定が保存されていない場合に表示されます。

Chapter 02 ルーターの基本操作

STEP6 設定の保存

今回利用するルーターはすべて設定のコマンドを入力するとただちに動作に反映しますが，不揮発性メモリーに保存されているわけではありません。そのため現在動作中の設定を保存することが必要です。実際のコマンドはそれぞれ以下のとおりです。

■ Cisco1841:

イネーブルモードにて，**copy** コマンドを用いて行います。

```
Router#copy running-config startup-config
Destination filename [startup-config]?
Building configuration...
[OK]
Router#
```

■ RTX1200:

管理ユーザーのアクセスレベルで **save** コマンドによって保存します。

```
#save
Saving ... CONFIG0 Done .
#
```

■ AX620R:

コンフィグモードにて，**write memory** コマンドによって行います。

```
Router(config)#write memory
Building configuration...
% Warning: do NOT enter CNTL/Z while saving to avoid config corruption.
Router(config)#
```

以上が各ルーターの基本的なオペレーションです。次章から実際の IPv6 に関する設定を行います。

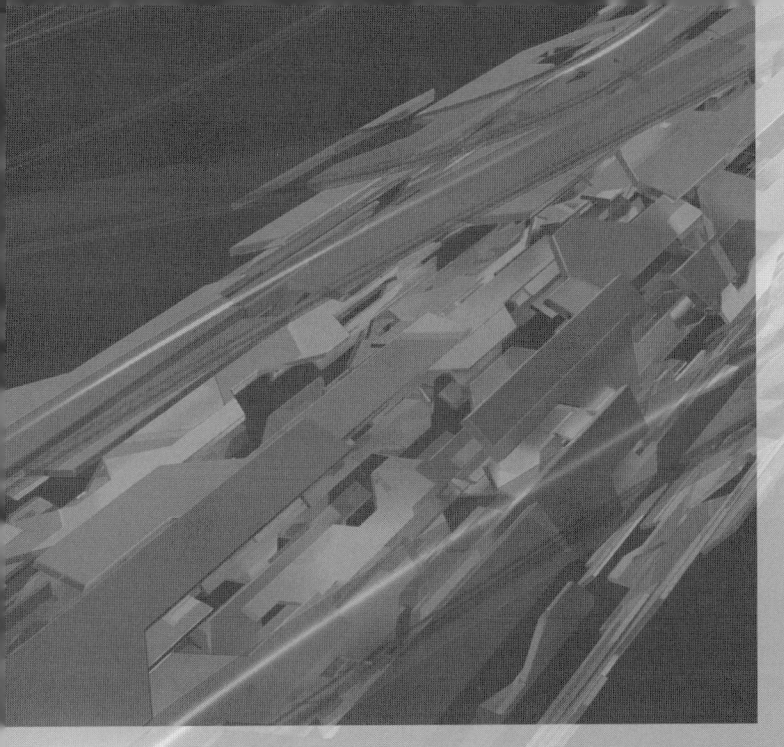

Chapter 03

IPv6 の基本的な設定

03-1 マルチベンダー機器
による実習

03-2 シスコ機器による実習

Chapter 03 IPv6の基本的な設定

この章では本書の実習において基本となる，3台のルーターによるトポロジーを作成し，IPv6の基本的な設定を実習します。本章における最終の設定はこのあとの実習の基本となる設定ですので，終了後に各ルーターにおいて設定を保存しておくとよいでしょう。

03-1 マルチベンダー機器による実習

この節のトポロジーは以下のとおりです。

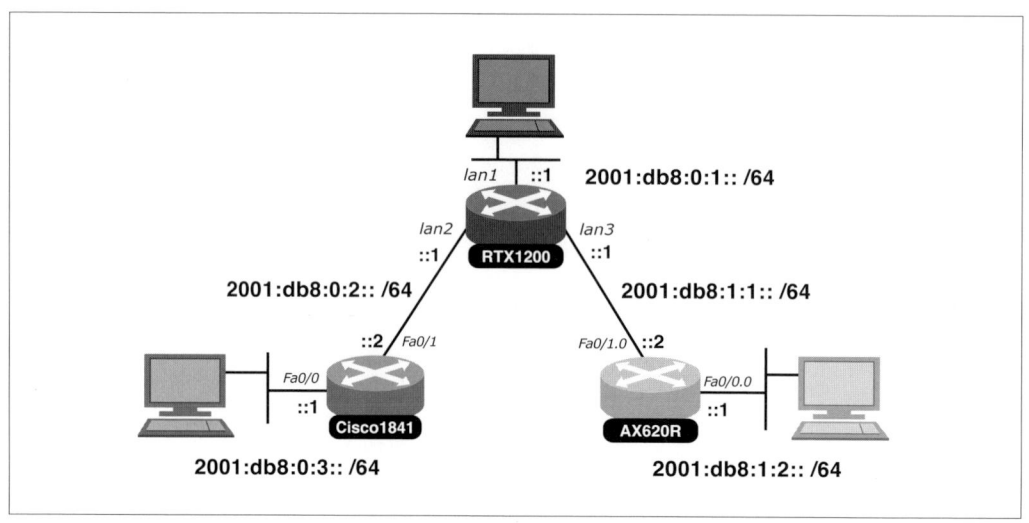

図3-1　ネットワークトポロジー

アドレスの割当は以下のとおりです。

表3-1　インターフェースアドレス割当一覧

ルーター	インターフェース	IPv6アドレス
Cisco1841	FastEthernet 0/0 (Fa0/0)	2001:db8:0:3::1/64
	FastEthernet 0/1 (Fa0/1)	2001:db8:0:2::2/64
RTX1200	lan1	2001:db8:0:1::1/64
	lan2	2001:db8:0:2::1/64
	lan3	2001:db8:1:1::1/64
AX620R	FastEthernet 0/0.0 (Fa0/0.0)	2001:db8:1:2::1/64
	FastEthernet 0/1.0 (Fa0/1.0)	2001:db8:1:1::2/64

STEP1 ルーターの初期化

各ルーターの設定に入る前に，事前準備を行います。まずはルーターの初期化です。

■ Cisco1841:

```
Router#erase startup-config
Router#reload
```

■ RTX1200:

次のコマンドで再起動すると、工場出荷時の設定で起動します。

```
#cold start
```

RTX1200は工場出荷時の設定ではLAN1にIPv6アドレスが割り当てられDHCPサーバーが起動しています。以下のコマンドでDHCPサーバーを停止し、IPv6アドレスの設定を削除します。

```
#no dhcp service server
#no ip lan1 address 192.168.100.1/24
```

■ AX620R:

```
Router(config)#erase startup-config
Router(config)#exit
Router#restart
```

次にホスト名の設定をします。

■ Cisco1841:

```
Router(config)#hostname Cisco1841
Cisco1841(config)#
```

■ RTX1200:

ホスト名の設定はできませんが、コマンドプロンプトの変更は可能です。

```
#console prompt RTX1200
RTX1200#
```

■ AX620R:

```
Router(config)#hostname AX620R
AX620R(config)#
```

シスコルーターはIPv6の有効化を行う必要があります。

■ Cisco1841:

```
Cisco1841(config)#ipv6 unicast-routing
```

STEP2 物理接続の実施

各ルーターおよびPCを"図3-1　ネットワークトポロジー"のとおりに接続します。

STEP3 各インターフェースへのIPv6アドレスの設定

各ルーターのそれぞれのインターフェースにIPv6アドレスを設定します。

■ Cisco1841:

```
Cisco1841(config)#interface fastEthernet 0/0
Cisco1841(config-if)#ipv6 enable
Cisco1841(config-if)#ipv6 address 2001:db8:0:3::1/64
Cisco1841(config-if)#no shutdown
Cisco1841(config-if)#exit
Cisco1841(config)#
Cisco1841(config)#interface fastEthernet 0/1
Cisco1841(config-if)#ipv6 enable
Cisco1841(config-if)#ipv6 address 2001:db8:0:2::2/64
Ciaco1841(config-if)#no shutdown
```

Chapter 03 IPv6の基本的な設定

■ RTX1200:

```
RTX1200#ipv6 lan1 address 2001:db8:0:1::1/64
RTX1200#ipv6 lan2 address 2001:db8:0:2::1/64
RTX1200#ipv6 lan3 address 2001:db8:1:1::1/64
```

■ AX620R:

```
AX620R(config)#interface FastEthernet0/0.0
AX620R(config-FastEthernet0/0.0)#ipv6 enable
AX620R(config-FastEthernet0/0.0)#ipv6 address 2001:db8:1:2::1/64
AX620R(config-FastEthernet0/0.0)#no shutdown
AX620R(config-FastEthernet0/0.0)#exit

AX620R(config)#interface FastEthernet0/1.0
AX620R(config-FastEthernet0/1.0)#ipv6 enable
AX620R(config-FastEthernet0/1.0)#ipv6 address 2001:db8:1:1::2/64
AX620R(config-FastEthernet0/1.0)#no shutdown
```

STEP4 ルーター相互の通信確認

各ルーターの対向のインターフェースに対して通信できるかを確認します。

■ Cisco1841:

対向の RTX1200 のインターフェースに対して **ping6** コマンドを実行します。

```
Cisco1841#ping6 2001:db8:0:2::1

Type escape sequence to abort.
Sending 5, 100-byte ICMP Echos to 2001:DB8:0:2::1, timeout is 2 seconds:
!!!!!
Success rate is 100 percent (5/5), round-trip min/avg/max = 0/0/0 ms
Cisco1841#
```

■ RTX1200:

対向の Cisco1841 のインターフェースに対して **ping6** コマンドを実行します。ping6 は打ち続けますので止めるときは Ctrl キーを押しながら C を押下して（以下 Ctrl+C）止めます。

```
RTX1200#ping6 2001:db8:0:2::2
received from 2001:db8:0:2::2, icmp_seq=0 hlim=64 time=0.697ms
received from 2001:db8:0:2::2, icmp_seq=1 hlim=64 time=0.697ms
received from 2001:db8:0:2::2, icmp_seq=2 hlim=64 time=0.682ms
received from 2001:db8:0:2::2, icmp_seq=3 hlim=64 time=0.707ms
received from 2001:db8:0:2::2, icmp_seq=4 hlim=64 time=0.697ms
received from 2001:db8:0:2::2, icmp_seq=5 hlim=64 time=0.679ms

6 packets transmitted, 6 packets received, 0.0% packet loss
round-trip min/avg/max = 0.679/0.693/0.707 ms
RTX1200#
```

続いて対向の AX620R に対して **ping6** コマンドを実行します。

```
RTX1200#ping6 2001:db8:1:1::2
received from 2001:db8:1:1::2, icmp_seq=0 hlim=64 time=0.449ms
received from 2001:db8:1:1::2, icmp_seq=1 hlim=64 time=0.417ms
received from 2001:db8:1:1::2, icmp_seq=2 hlim=64 time=0.414ms
received from 2001:db8:1:1::2, icmp_seq=3 hlim=64 time=0.417ms
received from 2001:db8:1:1::2, icmp_seq=4 hlim=64 time=0.415ms
```

```
received from 2001:db8:1:1::2, icmp_seq=5 hlim=64 time=0.417ms

6 packets transmitted, 6 packets received, 0.0% packet loss
round-trip min/avg/max = 0.414/0.421/0.449 ms
RTX1200#
```

■ AX620R:

対向の RTX1200 に対して **ping6** コマンドを実行します。

```
AX620R(config)#ping6 2001:db8:1:1::1
PING 2001:db8:1:1::2 > 2001:db8:1:1::1 56 data bytes
64 bytes from 2001:db8:1:1::1 icmp_seq=0 hlim=64 time=0.889 ms
64 bytes from 2001:db8:1:1::1 icmp_seq=1 hlim=64 time=0.553 ms
64 bytes from 2001:db8:1:1::1 icmp_seq=2 hlim=64 time=0.491 ms
64 bytes from 2001:db8:1:1::1 icmp_seq=3 hlim=64 time=0.484 ms
64 bytes from 2001:db8:1:1::1 icmp_seq=4 hlim=64 time=0.474 ms

--- 2001:db8:1:1::1 ping statistics ---
5 packets transmitted, 5 packets received, 0% packet loss
round-trip (ms)  min/avg/max = 0.474/0.578/0.889
AX620R(config)#
```

STEP5 RA の設定

IPv6 においてはホストがつながるインターフェースに対しては RA（Router Advertisement）を設定する必要があります。

■ Cisco1841:

設定は特に必要ありません。逆にルーター同士が接続しているインターフェースに対して RA の送信を停止します[※]。

```
Cisco1841(config)#interface fastEthernet 0/1
Cisco1841(config-if)#ipv6 nd ra suppress
Cisco1841(config-if)#
```

■ RTX1200:

PC がつながる LAN1 インターフェースに対して RA を設定します。まずプレフィックスを設定したのちに、そのプレフィックス番号をインターフェースに対して割り当てます。

```
RTX1200#ipv6 prefix 1 2001:db8:0:1::/64
RTX1200#ipv6 lan1 rtadv send 1
RTX1200#
```

■ AX620R:

PC がつながる Fa0/0.0 に対して RA を設定します。

```
AX620R(config)#interface FastEthernet0/0.0
AX620R(config-FastEthernet0/0.0)#ipv6 nd ra enable
AX620R(config-FastEthernet0/0.0)#
```

※1　IOS のバージョンによっては「ipv6 nd suppress-ra」とする必要があります。

Chapter 03 IPv6の基本的な設定

STEP6 DHCPv6 サービスの設定

　DHCPv6 サービスで提供する DNS サーバーの情報を定義したプロファイル "V6DHCP" を作成します。なおここでは DNS サーバーのアドレスを 2001:db8:2:1::53 とします。なお，RTX1200 では DHCPv6 サービスの提供はできません。また AX620R では DHCPv6 サービスを有効にする必要があります。

■ Cisco1841:

```
Cisco1841(config)#ipv6 dhcp pool V6DHCP
Cisco1841(config-dhcpv6)#dns-server 2001:db8:2:1::53
```

■ AX620R:

```
AX620R(config)#ipv6 dhcp enable
AX620R(config)#ipv6 dhcp server-profile v6dhcp
AX620R(config-ipv6-dhs-v6dhcp)#dns-server 2001:db8:2:1::53
```

STEP7 DHCPv6 サービスを動作させるインターフェースの設定

　DHCPv6 サービスを提供するインターフェースにおいて STEP6 で作成したプロファイル "V6DHCP" を割り当て，ステートフル構成フラグ（O フラグ）を有効にする必要があります。

■ Cisco1841:

```
Cisco1841(config)#interface FastEthernet0/0
Cisco1841(config-if)#ipv6 nd other-config-flag
Cisco1841(config-if)#ipv6 dhcp server V6DHCP
```

■ AX620R:

```
AX620R(config)#interface FastEthernet0/0.0
AX620R (config-FastEthernet0/0.0)#ipv6 dhcp server v6dhcp
AX620R (config-FastEthernet0/0.0)#ipv6 nd ra other-config-flag
```

　このとき AX620R は明示的に ipv6 nd ra enable として RA を動作させる必要があります。

STEP8 PC の動作確認

　設定したインターフェースに PC を接続し，IPv6 アドレスとともに IPv6 の DNS サーバーの情報が取得できているかを確認します。

```
Windows の場合：
    (ip アドレスの確認)      コマンドプロンプトから ipconfig /all
    (dns サーバーの確認)     コマンドプロンプトから nslookup
MacOSX の場合：
    (ip アドレスの確認)      シェルから ifconfig
    (dns サーバーの確認)     シェルから nslookup
```

STEP9　PCとルーターの通信確認

各ルーターにPCを接続し，IPv6のアドレスがRAにより取得できることと，通信ができることを確認します。

■ **Cisco1841:**

Fa0/0にPCを接続しIPv6アドレスを確認します。

次に2001:db8:0:3::1に対して **ping** コマンドを実行します。

```
Windowsの場合：
    (ipアドレスの確認)    コマンドプロンプトから ipconfig
    (pingコマンド)        コマンドプロンプトから ping -6 2001:db8:0:3::1
MacOSXの場合：
    (ipアドレスの確認)    ターミナルから ifconfig
    (pingコマンド)        ターミナルから ping6 2001:db8:0:3::1
```

■ **RTX1200:**

LAN1にPCを接続後，IPv6アドレスを確認します。2001:db8:0:1::1に対して **ping6** コマンドを実行します。

■ **AX620R:**

Fa0/0にPCを接続後，IPv6アドレスを確認します。2001:db8:1:2::1に対して **ping6** コマンドを実行します。

STEP10　IPv6アドレス割当状況の確認

各ルーターにおいてSTEP3で設定したIPv6アドレスが割り当てられていることを確認します。

■ **Cisco1841:**

show ipv6 interface brief コマンドで確認します。

```
Cisco1841#show ipv6 interface brief
FastEthernet0/0              [up/up]
    FE80::222:55FF:FE52:B7AE
    2001:DB8:0:3::1
FastEthernet0/1              [up/up]
    FE80::222:55FF:FE52:B7AF
    2001:DB8:0:2::2
Cisco1841#
```

■ **RTX1200:**

show ipv6 address コマンドで確認します。

```
RTX1200#show ipv6 address
LAN1 scope-id 1 [up]
  Received:     0 packet 0 octet
  Transmitted: 2 packets 156 octets

  global      2001:db8:0:1::1/64
  link-local  fe80::2a0:deff:fe37:80db/64
```

```
      link-local ff02::1/64
      link-local ff02::2/64
      link-local ff02::1:ff00:1/64
      link-local ff02::1:ff37:80db/64

    LAN2 scope-id 2 [up]
      Received:     25 packets 2432 octets
      Transmitted: 21 packets 2342 octets

      global     2001:db8:0:2::1/64
      link-local fe80::2a0:deff:fe37:80dc/64
      link-local ff02::1/64
      link-local ff02::2/64
      link-local ff02::1:ff00:1/64
      link-local ff02::1:ff37:80dc/64

    LAN3 scope-id 3 [up]
      Received:     32 packets 2736 octets
      Transmitted: 24 packets 2440 octets

      global     2001:db8:1:1::1/64
      link-local fe80::2a0:deff:fe37:80dd/64
      link-local ff02::1/64
      link-local ff02::2/64
      link-local ff02::1:ff00:1/64
      link-local ff02::1:ff37:80dd/64

    NULL scope-id 400 [up]
      Received:     0 packet 0 octet
      Transmitted: 0 packet 0 octet

      link-local fe80::1/64
      link-local ff02::1/64
      link-local ff02::2/64

    LOOPBACK1 scope-id 401 [up]
      Received:     0 packet 0 octet
      Transmitted: 0 packet 0 octet

      link-local fe80::1/64
      link-local ff02::1/64
      link-local ff02::2/64

    (中略)

    RTX1200#
```

■ **AX620R:**

show ipv6 address コマンドで確認します。

```
AX620R#show ipv6 address
Interface FastEthernet0/0.0 is up, line protocol is up
  Global address(es):
    2001:db8:1:2::1 prefixlen 64
    2001:db8:1:2::0 prefixlen 64 anycast
  Link-local address(es):
    fe80::260:b9ff:fe48:ccba prefixlen 64
    fe80::0 prefixlen 64 anycast
  Multicast address(es):
```

```
      ff02::1
      ff02::2
      ff02::1:ff00:0
      ff02::1:ff00:1
      ff02::1:ff48:ccba
Interface FastEthernet0/1.0 is up, line protocol is up
  Global address(es):
    2001:db8:1:1::2 prefixlen 64
    2001:db8:1:1::0 prefixlen 64 anycast
  Link-local address(es):
    fe80::260:b9ff:fe48:cc7a prefixlen 64
    fe80::0 prefixlen 64 anycast
  Multicast address(es):
    ff02::1
    ff02::2
    ff02::1:ff00:0
    ff02::1:ff00:2
    ff02::1:ff48:cc7a
Interface Loopback0.0 is up, line protocol is up
  Orphan address(es):
    ::1 prefixlen 128
Interface Loopback1.0 is up, line protocol is up
Interface Null0.0 is up, line protocol is up
Interface Null1.0 is up, line protocol is up
AX620R#
```

STEP11　DHCPv6 サービスの動作確認

各ルーターにおいて STEP6 および 7 で設定した DHCPv6 サービスが動作していることを確認します。

■ Cisco1841:

show ipv6 dhcp pool コマンドで確認します。

```
Cisco1841#show ipv6 dhcp pool
DHCPv6 pool: V6DHCP
  DNS server: 2001:DB8:2:1::53
  Active clients: 0

Cisco1841#show ipv6 dhcp interface
FastEthernet0/0 is in server mode
  Using pool: V6DHCP
  Preference value: 0
  Hint from client: ignored
  Rapid-Commit: disabled
Cisco1841#
```

■ AX620R:

show ipv6 dhcp server コマンドで確認します。

```
AX620R(config)#show ipv6 dhcp server
DHCPv6 server is enabled
  System DUID 00:03:00:01:00:30:13:f6:1e:8e
  Statistics:
    Information request-reply:
      1 receive, 0 drops, 104 seconds ago
      1 send, 104 seconds ago
Interface FastEthernet0/0.0 is active
```

```
     Neighbor fe80::cc6a:d38f:b8fa:a020 is active
       Client identifier 00:01:00:01:0f:eb:90:dd:00:1e:c9:09:16:03
       DNS Servers:
         2001:db8:2:1::53
     Statistics:
       Information request-reply:
         1 receive, 0 drops, 105 seconds ago
         1 send, 105 seconds ago
AX620R(config)#
```

■実習終了時設定内容

■ Cisco1841:

```
Cisco1841#show run
Building configuration...

Current configuration : 1103 bytes
!
version 12.4
service timestamps debug datetime msec
service timestamps log datetime msec
no service password-encryption
!
hostname Cisco1841
!
boot-start-marker
boot-end-marker
!
logging message-counter syslog
!
no aaa new-model
dot11 syslog
ip source-route
!
ip cef
ipv6 unicast-routing
ipv6 cef
ipv6 dhcp pool V6DHCP
 dns-server 2001:DB8:2:1::53
!
multilink bundle-name authenticated
!
archive
 log config
  hidekeys
!
interface FastEthernet0/0
 no ip address
 duplex auto
 speed auto
 ipv6 address 2001:DB8:0:3::1/64
 ipv6 enable
 ipv6 nd other-config-flag
 ipv6 dhcp server V6DHCP
!
interface FastEthernet0/1
 no ip address
 duplex auto
```

```
 speed auto
 ipv6 address 2001:DB8:0:2::2/64
 ipv6 enable
 ipv6 nd ra suppress
!
interface Serial0/0/0
 no ip address
 shutdown
 no fair-queue
 clock rate 125000
!
interface Serial0/0/1
 no ip address
 shutdown
 clock rate 125000
!
ip forward-protocol nd
no ip http server
no ip http secure-server
!
control-plane
!
line con 0
line aux 0
line vty 0 4
 login
!
scheduler allocate 20000 1000
end

Cisco1841#
```

■RTX1200：

```
RTX1200#show config
# RTX1200 Rev.10.01.29 (Wed Feb  9 17:21:33 2011)
# MAC Address : 00:a0:de:37:80:db, 00:a0:de:37:80:dc, 00:a0:de:37:80:dd
# Memory 128Mbytes, 3LAN, 1BRI
# main:   RTX1200 ver=b0 serial=D26007524 MAC-Address=00:a0:de:37:80:db MAC-Addr
ess=00:a0:de:37:80:dc MAC-Address=00:a0:de:37:80:dd
# Reporting Date: Feb 20 16:51:50 2013
console character ascii
console prompt RTX1200
ipv6 prefix 1 2001:db8:0:1::/64
ipv6 lan1 address 2001:db8:0:1::1/64
ipv6 lan1 rtadv send 1
ipv6 lan2 address 2001:db8:0:2::1/64
ipv6 lan3 address 2001:db8:1:1::1/64
RTX1200#
```

■AX620R：

```
AX620R(config)#show run
Current configuration : 1106 bytes

! NEC Portable Internetwork Core Operating System Software
! IX Series IX2010 (magellan-sec) Software, Version 8.3.13, RELEASE SOFTWARE
! Compiled Feb 25-Wed-2009 13:20:17 JST #1
! Current time Feb 20-Wed-2013 16:51:40 JST
!
```

```
hostname AX620R
timezone +09 00
!
ipv6 dhcp enable
!
ipv6 dhcp server-profile v6dhcp
   dns-server 2001:db8:2:1::53
!
device FastEthernet0/0
!
device FastEthernet0/1
!
device FastEthernet1/0
!
device BRI1/0
   isdn switch-type hsd128k
!
interface FastEthernet0/0.0
  no ip address
  ipv6 enable
  ipv6 address 2001:db8:1:2::1/64
  ipv6 dhcp server v6dhcp
  ipv6 nd ra enable
  ipv6 nd ra other-config-flag
  no shutdown
!
interface FastEthernet0/1.0
  no ip address
  ipv6 enable
  ipv6 address 2001:db8:1:1::2/64
  no shutdown
!
interface FastEthernet1/0.0
  no ip address
  shutdown
!
interface BRI1/0.0
   encapsulation ppp
   no auto-connect
   no ip address
   shutdown
!
interface Loopback0.0
   no ip address
!
interface Null0.0
   no ip address
AX620R(config)#
```

03-2 シスコ機器による実習

ここでは3つのシスコルーターを使ったスタティックルートによるネットワークを構成します。トポロジーは以下のとおりです。

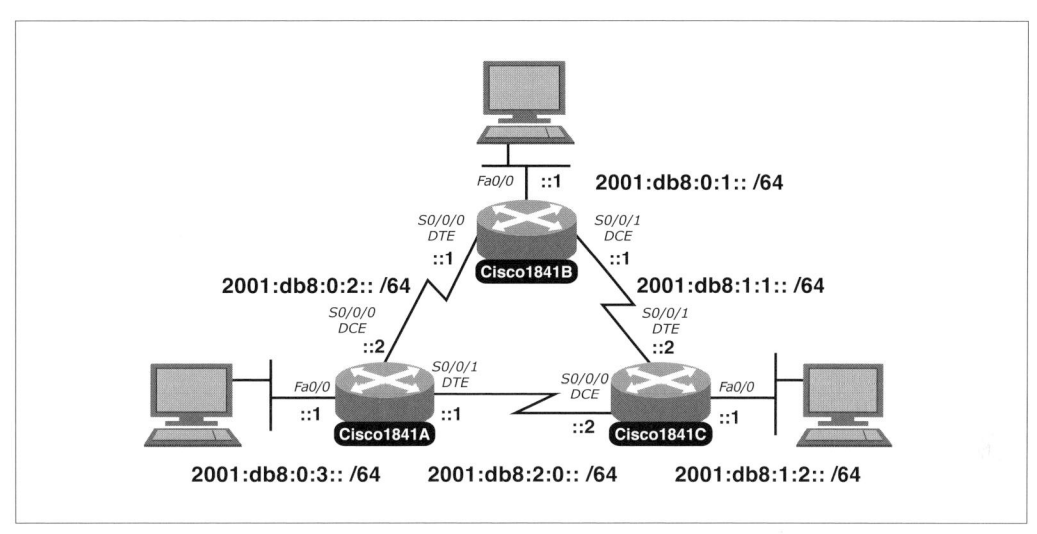

図3-2 ネットワークトポロジー

アドレスの割当は以下のとおりです。

表3-2 インターフェースアドレス割当一覧

ルーター	インターフェース	IPv6アドレス
Cisco1841A	FastEthernet 0/0 (Fa0/0)	2001:db8:0:3::1/64
	Serial 0/0/0 (S0/0/0)	2001:db8:0:2::2/64
	Serial 0/0/1 (S0/0/1)	2001:db8:2:0::1/64
Cisco1841B	FastEthernet 0/0 (Fa0/0)	2001:db8:0:1::1/64
	Serial 0/0/0 (S0/0/0)	2001:db8:0:2::1/64
	Serial 0/0/1 (S0/0/1)	2001:db8:1:1::1/64
Cisco1841C	FastEthernet 0/0 (Fa0/0)	2001:db8:1:2::1/64
	Serial 0/0/0 (S0/0/0)	2001:db8:2:0::2/64
	Serial 0/0/1 (S0/0/1)	2001:db8:1:1::2/64

STEP1 ルーターの初期化

各ルーターの設定に入る前に，事前準備を行います。まずはルーターの初期化です。

```
Router#erase startup-config
Router#reload
```

次にホスト名の設定をします。下記はCisco1841Aの例ですが，他ルーターも同様に行います。

```
Router(config)#hostname Cisco1841A
Cisco1841A(config)#
```

Chapter 03 IPv6 の基本的な設定

次に各ルーターで IPv6 の有効化を行います。

```
Cisco1841A(config)#ipv6 unicast-routing
```

STEP2　物理接続の実施

各ルーターおよび PC を "図 3-2　ネットワークトポロジー" のとおりに接続します。

STEP3　各インターフェースへの IPv6 の設定

トポロジーのように IPv6 アドレスを設定します。

■ Cisco1841A：

```
Cisco1841A(config)#interface fastethernet 0/0
Cisco1841A(config-if)#ipv6 enable
Cisco1841A(config-if)#ipv6 address 2001:db8:0:3::1/64
Cisco1841A(config-if)#no shutdown
Cisco1841A(config-if)#exit

Cisco1841A(config)#interface serial 0/0/0
Cisco1841A(config-if)#clock rate 125000
Cisco1841A(config-if)#ipv6 enable
Cisco1841A(config-if)#ipv6 address 2001:db8:0:2::2/64
Ciaco1841A(config-if)#no shutdown

Cisco1841A(config)#interface serial 0/0/1
Cisco1841A(config-if)#ipv6 enable
Cisco1841A(config-if)#ipv6 address 2001:db8:2::1/64
Ciaco1841A(config-if)#no shutdown
```

■ Cisco1841B：

```
Cisco1841B(config)#interface fastethernet 0/0
Cisco1841B(config-if)#ipv6 enable
Cisco1841B(config-if)#ipv6 address 2001:db8:0:1::1/64
Cisco1841B(config-if)#no shutdown
Cisco1841B(config-if)#exit

Cisco1841B(config)#interface serial 0/0/0
Cisco1841B(config-if)#ipv6 enable
Cisco1841B(config-if)#ipv6 address 2001:db8:0:2::1/64
Ciaco1841B(config-if)#no shutdown

Cisco1841B(config)#interface serial 0/0/1
Cisco1841B(config-if)#clock rate 125000
Cisco1841B(config-if)#ipv6 enable
Cisco1841B(config-if)#ipv6 address 2001:db8:1:1::1/64
Ciaco1841B(config-if)#no shutdown
```

■ Cisco1841C：

```
Cisco1841C(config)#interface fastethernet 0/0
Cisco1841C(config-if)#ipv6 enable
Cisco1841C(config-if)#ipv6 address 2001:db8:1:2::1/64
Cisco1841C(config-if)#no shutdown
Cisco1841C(config-if)#exit

Cisco1841C(config)#interface serial 0/0/0
```

```
Cisco1841C(config-if)#clock rate 125000
Cisco1841C(config-if)#ipv6 enable
Cisco1841C(config-if)#ipv6 address 2001:db8:2::2/64
Ciaco1841C(config-if)#no shutdown

Cisco1841C(config)#interface serial 0/0/1
Cisco1841C(config-if)#ipv6 enable
Cisco1841C(config-if)#ipv6 address 2001:db8:1:1::2/64
Ciaco1841C(config-if)#no shutdown
```

STEP4 ルーター相互の通信確認

各ルーターの対向のインターフェースに対して通信できるかを確認します。

■ Cisco1841A:

対向のCisco1841Bのインターフェースに対して**ping**コマンドを実行して、通信を確認します。

```
Cisco1841A#ping 2001:db8:0:2::1

Type escape sequence to abort.
Sending 5, 100-byte ICMP Echos to 2001:DB8:0:2::1, timeout is 2 seconds:
!!!!!
Success rate is 100 percent (5/5), round-trip min/avg/max = 0/0/0 ms
Cisco1841A#
```

同じく対向のCisco1841Cのインターフェースに対して**ping**コマンドを実行します。

```
Cisco1841A#ping 2001:db8:2::2

Type escape sequence to abort.
Sending 5, 100-byte ICMP Echos to 2001:DB8:2::2, timeout is 2 seconds:
!!!!!
Success rate is 100 percent (5/5), round-trip min/avg/max = 12/15/16 ms
Cisco1841A#
```

Cisco1841B, Cisco1841Cについてもトポロジーを確認して、対向ルーターのインターフェースに対して**ping**コマンドを実行して、通信を確認します。

■ Cisco1841B:

```
Cisco1841B#ping 2001:db8:0:2::2

Type escape sequence to abort.
Sending 5, 100-byte ICMP Echos to 2001:DB8:0:2::2, timeout is 2 seconds:
!!!!!
Success rate is 100 percent (5/5), round-trip min/avg/max = 12/14/16 ms
Cisco1841B#ping 2001:db8:1:1::2

Type escape sequence to abort.
Sending 5, 100-byte ICMP Echos to 2001:DB8:1:1::2, timeout is 2 seconds:
!!!!!
Success rate is 100 percent (5/5), round-trip min/avg/max = 12/13/16 ms
Cisco1841B#
```

■ Cisco1841C:

```
Cisco1841C#ping 2001:db8:0:2::1
```

Chapter 03　IPv6の基本的な設定

```
Type escape sequence to abort.
Sending 5, 100-byte ICMP Echos to 2001:DB8:2::1, timeout is 2 seconds:
!!!!!
Success rate is 100 percent (5/5), round-trip min/avg/max = 12/15/16 ms
Cisco1841C#ping 2001:db8:1:1::1

Type escape sequence to abort.
Sending 5, 100-byte ICMP Echos to 2001:DB8:1:1::1, timeout is 2 seconds:
!!!!!
Success rate is 100 percent (5/5), round-trip min/avg/max = 12/14/16 ms
Cisco1841C#
```

STEP5　RAの設定

IPv6においてはホストがつながるインターフェースに対してはRA (Router Advertisement) を設定する必要があります。

しかし，シスコルーターにおいてはデフォルトでRAを送出する設定になっているため，設定は特に必要ありません。逆にルーター同士が接続しているインターフェースに対してRAの送信を停止します。下記の例はCisco1841AですがCisco1841B, Cisco1841Cについても同様に設定します。[※2]

```
Cisco1841A(config)#interface serial 0/0/0
Cisco1841A(config-if)#ipv6 nd ra suppress
Cisco1841A(config-if)#

Cisco1841A(config)#interface serial 0/0/1
Cisco1841A(config-if)#ipv6 nd ra suppress
Cisco1841A(config-if)#
```

STEP6　DHCPv6サービスの設定

DHCPv6で提供するDNSサーバー情報を定義したプロファイル"V6DHCP"を作成します。なおここではDNSサーバーのアドレスを2001:db8:2:1::53とします。下記の例はCisco1841Aですが，Cisco1841B, Cisco1841Cについても同様に設定します。

■ Cisco1841A：

```
Cisco1841A(config)#ipv6 dhcp pool V6DHCP
Cisco1841A(config-dhcpv6)#dns-server 2001:db8:2:1::53
```

STEP7　DHCPv6サービスを動作させるインターフェースの設定

DHCPv6を動作させるインターフェースにおいてSTEP6で作成したプロファイル"V6DHCP"を割り当てます。また，同時にステートフル構成フラグ（Oフラグ）を有効にする必要があります。ここでも同じくCisco1841Aの例のみ記載します。

■ Cisco1841A：

```
Cisco1841A(config)#interface FastEthernet0/0
Cisco1841A(config-if)#ipv6 nd other-config-flag
Cisco1841A(config-if)#ipv6 dhcp server V6DHCP
```

※2　古いバージョンのIOSでは「ipv6 nd suppress-ra」と入力する必要がある場合があります。

STEP8　PCの動作確認

設定したインターフェースにPCを接続し、IPv6アドレスとともにIPv6のDHCPv6サーバーの情報が取得できているかを確認します。

```
Windowsの場合：
    (ipアドレスの確認)      コマンドプロンプトから ipconfig
    (pingコマンド)          コマンドプロンプトから nslookup
MacOSXの場合：
    (ipアドレスの確認)      シェルから ifconfig
    (pingコマンド)          シェルから nslookup
```

STEP9　PCとルーターの通信確認

各ルーターにPCを接続し、IPv6のアドレスがRAにより取得できることと、通信ができることを確認します。

■ Cisco1841A：

Fa0/0にPCを接続しIPv6アドレスを確認します。

次に 2001:db8:0:3::1 に対して **ping** コマンドを実行して確認します。

```
Windowsの場合：
    (ipアドレスの確認)      コマンドプロンプトから ipconfig
    (pingコマンド)          コマンドプロンプトから ping -6 2001:db8:0:3::1
MacOSXの場合：
    (ipアドレスの確認)      シェルから ifconfig
    (pingコマンド)          シェルから ping6 2001:db8:0:3::1
```

STEP10　IPv6アドレス割当状況の確認

各ルーターにおいてSTEP3で設定したIPv6アドレスが割り当てられていることを確認します。
以下にCisco1841Aの例を示しますがCisco1841BとCisco1841Cも同時に確認してください。

■ Cisco1841A：

show ipv6 interface brief コマンドで確認します。

```
Cisco1841A#show ipv6 interface brief
FastEthernet0/0            [up/up]
    FE80::223:EBFF:FE44:ED0C
    2001:DB8:0:3::1
FastEthernet0/1            [administratively down/down]
    unassigned
Serial0/0/0                [up/up]
    FE80::223:EBFF:FE44:ED0C
    2001:DB8:0:2::2
Serial0/0/1                [up/up]
    FE80::223:EBFF:FE44:ED0C
```

Chapter 03 IPv6の基本的な設定

```
    2001:DB8:2::1
Cisco1841A#
```

■ Cisco1841B：

show ipv6 interface brief コマンドで確認します。

```
Cisco1841B#show ipv6 interface brief
FastEthernet0/0            [up/up]
    FE80::222:55FF:FE52:B23C
    2001:DB8:0:1::1
FastEthernet0/1            [administratively down/down]
Serial0/0/0                [up/up]
    FE80::222:55FF:FE52:B23C
    2001:DB8:0:2::1
Serial0/0/1                [up/up]
    FE80::222:55FF:FE52:B23C
    2001:DB8:1:1::1
Cisco1841B#
```

■ Cisco1841C：

show ipv6 interface brief コマンドで確認します。

```
Cisco1841C#show ipv6 interface brief
FastEthernet0/0            [up/up]
    FE80::222:55FF:FE52:B7AE
    2001:DB8:1:2::1
FastEthernet0/1            [administratively down/down]
Serial0/0/0                [up/up]
    FE80::222:55FF:FE52:B7AE
    2001:DB8:2::2
Serial0/0/1                [up/up]
    FE80::222:55FF:FE52:B7AE
    2001:DB8:1:1::2
Cisco1841C#
```

STEP11　DHCPv6 サービスの動作確認

各ルーターにおいて STEP6 および 7 で設定した DHCPv6 サービスが動作していることを確認します。下記の例は Cisco1841A ですが，Cisco1841B，Cisco1841C についても同様に確認します。

■ Cisco1841A：

```
Cisco1841A#show ipv6 dhcp pool
DHCPv6 pool: V6DHCP
  DNS server: 2001:DB8:2:1::53
  Active clients: 0

Cisco1841A#show ipv6 dhcp interface
FastEthernet0/0 is in server mode
  Using pool: V6DHCP
  Preference value: 0
  Hint from client: ignored
  Rapid-Commit: disabled
Cisco1841A#
```

■実習終了時設定内容

■Cisco1841A：

```
Cisco1841A#show run
Building configuration...

Current configuration : 1069 bytes
!
version 12.4
service timestamps debug datetime msec
service timestamps log datetime msec
no service password-encryption
!
hostname Cisco1841A
!
boot-start-marker
boot-end-marker
!
logging message-counter syslog
!
no aaa new-model
dot11 syslog
ip source-route
!
ip cef
ipv6 unicast-routing
ipv6 cef
ipv6 dhcp pool V6DHCP
 dns-server 2001:DB8:2:1::53
!
multilink bundle-name authenticated
!
archive
 log config
  hidekeys
!
interface FastEthernet0/0
 no ip address
 duplex auto
 speed auto
 ipv6 address 2001:DB8:0:3::1/64
 ipv6 enable
 ipv6 nd other-config-flag
 ipv6 dhcp server V6DHCP
!
interface FastEthernet0/1
 no ip address
 shutdown
 duplex auto
 speed auto
!
interface Serial0/0/0
 no ip address
 ipv6 address 2001:DB8:0:2::2/64
 ipv6 enable
 clock rate 125000
!
interface Serial0/0/1
 no ip address
```

```
 ipv6 address 2001:DB8:2::1/64
 ipv6 enable
!
ip forward-protocol nd
no ip http server
no ip http secure-server
!
control-plane
!
line con 0
line aux 0
line vty 0 4
 login
!
scheduler allocate 20000 1000
end
```

■ **Cisco1841B：**

```
Cisco1841B#show run
Building configuration...

Current configuration : 1085 bytes
!
version 12.4
service timestamps debug datetime msec
service timestamps log datetime msec
no service password-encryption
!
hostname Cisco1841B
!
boot-start-marker
boot-end-marker
!
logging message-counter syslog
!
no aaa new-model
dot11 syslog
ip source-route
!
ip cef
ipv6 unicast-routing
ipv6 cef
ipv6 dhcp pool V6DHCP
 dns-server 2001:DB8:2:1::53
!
multilink bundle-name authenticated
!
archive
 log config
  hidekeys
!
interface FastEthernet0/0
 no ip address
 duplex auto
 speed auto
 ipv6 address 2001:DB8:0:1::1/64
 ipv6 enable
 ipv6 nd other-config-flag
 ipv6 dhcp server V6DHCP
```

```
!
interface FastEthernet0/1
 no ip address
 shutdown
 duplex auto
 speed auto
!
interface Serial0/0/0
 no ip address
 ipv6 address 2001:DB8:0:2::1/64
 ipv6 enable
!
interface Serial0/0/1
 no ip address
 ipv6 address 2001:DB8:1:1::1/64
 ipv6 enable
 clock rate 125000
!
ip forward-protocol nd
no ip http server
no ip http secure-server
!
control-plane
!
line con 0
line aux 0
line vty 0 4
 login
!
scheduler allocate 20000 1000
end
```

■ Cisco1841C：

```
Cisco1841C#show run
Building configuration...

Current configuration : 1083 bytes
!
version 12.4
service timestamps debug datetime msec
service timestamps log datetime msec
no service password-encryption
!
hostname Cisco1841C
!
boot-start-marker
boot-end-marker
!
logging message-counter syslog
!
no aaa new-model
dot11 syslog
ip source-route
!
!
!
!
ip cef
ipv6 unicast-routing
```

```
ipv6 cef
ipv6 dhcp pool V6DHCP
 dns-server 2001:DB8:2:1::53
!
multilink bundle-name authenticated
!
archive
 log config
  hidekeys
!
interface FastEthernet0/0
 no ip address
 duplex auto
 speed auto
 ipv6 address 2001:DB8:1:2::1/64
 ipv6 enable
 ipv6 nd other-config-flag
 ipv6 dhcp server V6DHCP
!
interface FastEthernet0/1
 no ip address
 shutdown
 duplex auto
 speed auto
!
interface Serial0/0/0
 no ip address
 ipv6 address 2001:DB8:2::2/64
 ipv6 enable
 clock rate 125000
!
interface Serial0/0/1
 no ip address
 ipv6 address 2001:DB8:1:1::2/64
 ipv6 enable
!
ip forward-protocol nd
no ip http server
no ip http secure-server
!
control-plane
!
line con 0
line aux 0
line vty 0 4
 login
!
scheduler allocate 20000 1000
end
```

Chapter 04

スタティックルートの設定

04-1 マルチベンダー機器
　　 による実習

04-2 シスコ機器による実習

Chapter 04 スタティックルートの設定

この章では前章に続いて，各ルーターにスタティックルートを設定することで，直接つながっていないホストやルーター同士の接続を可能にするように設定していきます。その手法として，まずはスタティックルートを設定することを学びます。

04-1 マルチベンダー機器による実習

第3章と同じトポロジーを用いて学習します。実習は第3章の終了時点の状態から始めます。

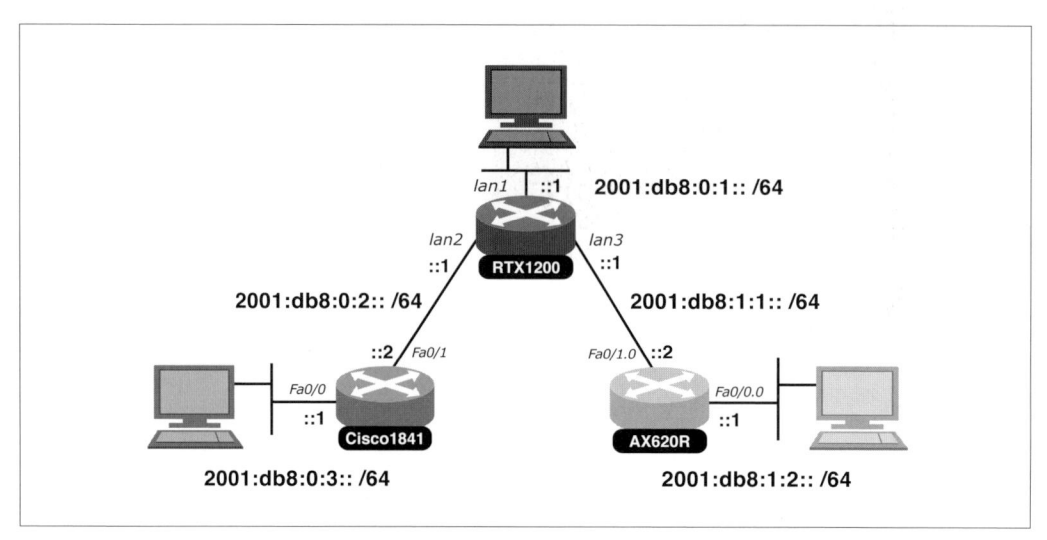

図4-1　ネットワークトポロジー

アドレスの割当は以下のとおりです。

表4-1　インターフェースアドレス割当一覧

ルーター	インターフェース	IPv6アドレス
Cisco1841	FastEthernet 0/0 (Fa0/0)	2001:db8:0:3::1/64
	FastEthernet 0/1 (Fa0/1)	2001:db8:0:2::2/64
RTX1200	lan1	2001:db8:0:1::1/64
	lan2	2001:db8:0:2::1/64
	lan3	2001:db8:1:1::1/64
AX620R	FastEthernet 0/0.0 (Fa0/0.0)	2001:db8:1:2::1/64
	FastEthernet 0/1.0 (Fa0/1.0)	2001:db8:1:1::2/64

STEP1　ルーティングテーブルの確認

各ルーターのルーティングテーブルを確認します。

■ Cisco1841:

show ipv6 route コマンドで確認します。

```
Cisco1841#show ipv6 route
IPv6 Routing Table - 6 entries
```

```
Codes: C - Connected, L - Local, S - Static, R - RIP, B - BGP
       U - Per-user Static route
       I1 - ISIS L1, I2 - ISIS L2, IA - ISIS interarea, IS - ISIS summary
       O - OSPF intra, OI - OSPF inter, OE1 - OSPF ext 1, OE2 - OSPF ext 2
       ON1 - OSPF NSSA ext 1, ON2 - OSPF NSSA ext 2
C   2001:DB8:0:2::/64 [0/0]
     via ::, FastEthernet0/1
L   2001:DB8:0:2::2/128 [0/0]
     via ::, FastEthernet0/1
C   2001:DB8:0:3::/64 [0/0]
     via ::, FastEthernet0/0
L   2001:DB8:0:3::1/128 [0/0]
     via ::, FastEthernet0/0
L   FE80::/10 [0/0]
     via ::, Null0
L   FF00::/8 [0/0]
     via ::, Null0
Cisco1841#
```

■ RTX1200：

show ipv6 route コマンドで確認します。

```
RTX1200#show ipv6 route
Destination              Gateway              Interface   Type
2001:db8:0:1::/64        -                    LAN1        implicit
2001:db8:0:2::/64        -                    LAN2        implicit
2001:db8:1:1::/64        -                    LAN3        implicit
RTX1200#
```

■ AX620R：

show ipv6 route コマンドで確認します。

```
AX620R#show ipv6 route
IPv6 Routing Table - 6 entries, unlimited
Codes: C - Connected, L - Local, S - Static
       R - RIPng, O - OSPF, IA - OSPF inter area
       E1 - OSPF external type 1, E2 - OSPF external type 2, B - BGP
       s - Summary
Timers: Uptime/Age
C     2001:db8:1:1::/64 global [0/1]
        via ::, FastEthernet0/1.0, 2:28:28/0:00:00
L     2001:db8:1:1::/128 global [0/1]
        via ::, FastEthernet0/1.0, 2:28:29/0:00:00
L     2001:db8:1:1::2/128 global [0/1]
        via ::, FastEthernet0/1.0, 2:28:28/0:00:00
C     2001:db8:1:2::/64 global [0/1]
        via ::, FastEthernet0/0.0, 0:05:49/0:00:00
L     2001:db8:1:2::/128 global [0/1]
        via ::, FastEthernet0/0.0, 0:05:50/0:00:00
L     2001:db8:1:2::1/128 global [0/1]
        via ::, FastEthernet0/0.0, 0:05:49/0:00:00
AX620R#
```

Chapter 04 スタティックルートの設定

STEP2 スタティックルートの設定

各ルーターにスタティックルートを設定します。

■ Cisco1841:

```
Cisco1841(config)#ipv6 route 2001:db8:0:1::/64 2001:db8:0:2::1
Cisco1841(config)#ipv6 route 2001:db8:1:1::/64 2001:db8:0:2::1
Cisco1841(config)#ipv6 route 2001:db8:1:2::/64 2001:db8:0:2::1
```

■ RTX1200:

```
RTX1200#ipv6 route 2001:db8:0:3::/64 gateway 2001:db8:0:2::2%2
RTX1200#ipv6 route 2001:db8:1:2::/64 gateway 2001:db8:1:1::2%3
```

■ AX620R:

```
AX620R(config)#ipv6 route 2001:db8:0:3::/64 2001:db8:1:1::1
AX620R(config)#ipv6 route 2001:db8:0:2::/64 2001:db8:1:1::1
AX620R(config)#ipv6 route 2001:db8:0:1::/64 2001:db8:1:1::1
```

STEP3 ルーティングテーブルの確認

各ルーターのルーティングテーブルを確認します。

■ Cisco:1841:

show ipv6 route コマンドで確認します。

```
Cisco1841#show ipv6 route
IPv6 Routing Table - 9 entries
Codes: C - Connected, L - Local, S - Static, R - RIP, B - BGP
       U - Per-user Static route
       I1 - ISIS L1, I2 - ISIS L2, IA - ISIS interarea, IS - ISIS summary
       O - OSPF intra, OI - OSPF inter, OE1 - OSPF ext 1, OE2 - OSPF ext 2
       ON1 - OSPF NSSA ext 1, ON2 - OSPF NSSA ext 2
S   2001:DB8:0:1::/64 [1/0]
     via 2001:DB8:0:2::1
C   2001:DB8:0:2::/64 [0/0]
     via ::, FastEthernet0/1
L   2001:DB8:0:2::2/128 [0/0]
     via ::, FastEthernet0/1
C   2001:DB8:0:3::/64 [0/0]
     via ::, FastEthernet0/0
L   2001:DB8:0:3::1/128 [0/0]
     via ::, FastEthernet0/0
S   2001:DB8:1:1::/64 [1/0]
     via 2001:DB8:0:2::1
S   2001:DB8:1:2::/64 [1/0]
     via 2001:DB8:0:2::1
L   FE80::/10 [0/0]
     via ::, Null0
L   FF00::/8 [0/0]
     via ::, Null0
Cisco1841#
```

■ RTX1200:

show ipv6 route コマンドで確認します。

```
RTX1200#show ipv6 route
Destination                    Gateway                    Interface   Type
```

```
2001:db8:0:1::/64          -                    LAN1     implicit
2001:db8:0:2::/64          -                    LAN2     implicit
2001:db8:0:3::/64          2001:db8:0:2::2      LAN2     static
2001:db8:1:1::/64          -                    LAN3     implicit
2001:db8:1:2::/64          2001:db8:1:1::2      LAN3     static
RTX1200#
```

■ AX620R:

show ipv6 route コマンドで確認します。

```
AX620R(config)#show ipv6 route
IPv6 Routing Table - 9 entries, unlimited
Codes: C - Connected, L - Local, S - Static
       R - RIPng, O - OSPF, IA - OSPF inter area
       E1 - OSPF external type 1, E2 - OSPF external type 2, B - BGP
       s - Summary
Timers: Uptime/Age
S       2001:db8:0:1::/64 global [1/1]
            via 2001:db8:1:1::1, FastEthernet0/1.0, 0:01:02/0:00:00
S       2001:db8:0:2::/64 global [1/1]
            via 2001:db8:1:1::1, FastEthernet0/1.0, 0:01:07/0:00:00
S       2001:db8:0:3::/64 global [1/1]
            via 2001:db8:1:1::1, FastEthernet0/1.0, 0:02:55/0:00:00
C       2001:db8:1:1::/64 global [0/1]
            via ::, FastEthernet0/1.0, 0:24:17/0:00:00
L       2001:db8:1:1::/128 global [0/1]
            via ::, FastEthernet0/1.0, 0:24:18/0:00:00
L       2001:db8:1:1::2/128 global [0/1]
            via ::, FastEthernet0/1.0, 0:24:17/0:00:00
C       2001:db8:1:2::/64 global [0/1]
            via ::, FastEthernet0/0.0, 0:23:41/0:00:00
L       2001:db8:1:2::/128 global [0/1]
            via ::, FastEthernet0/0.0, 0:23:43/0:00:00
L       2001:db8:1:2::1/128 global [0/1]
            via ::, FastEthernet0/0.0, 0:23:43/0:00:00
AX620R(config)#
```

STEP4　PC からルーターまでの通信確認

各 PC からそれぞれが接続しているルーターのインターフェースまでの通信を確認します。

■ Cisco1841:

接続した PC から以下の IPv6 アドレスに対して **ping** コマンドを実行します。

```
2001:db8:0:1::1
2001:db8:1:2::1
```

■ RTX1200:

接続した PC から以下の IPv6 アドレスに対して **ping** コマンドを実行します。

```
2001:db8:0:3::1
2001:db8:1:2::1
```

■ AX620R:

接続した PC から以下の IPv6 アドレスに対して **ping** コマンドを実行します。

```
2001:db8:0:1::1
2001:db8:0:3::1
```

もし応答がなかった場合は設定を確認してください。

04-2 シスコ機器による実習

シスコルーターも第3章と同じトポロジーを用いて学習します。実習は第3章の終了時点の状態から始めます。

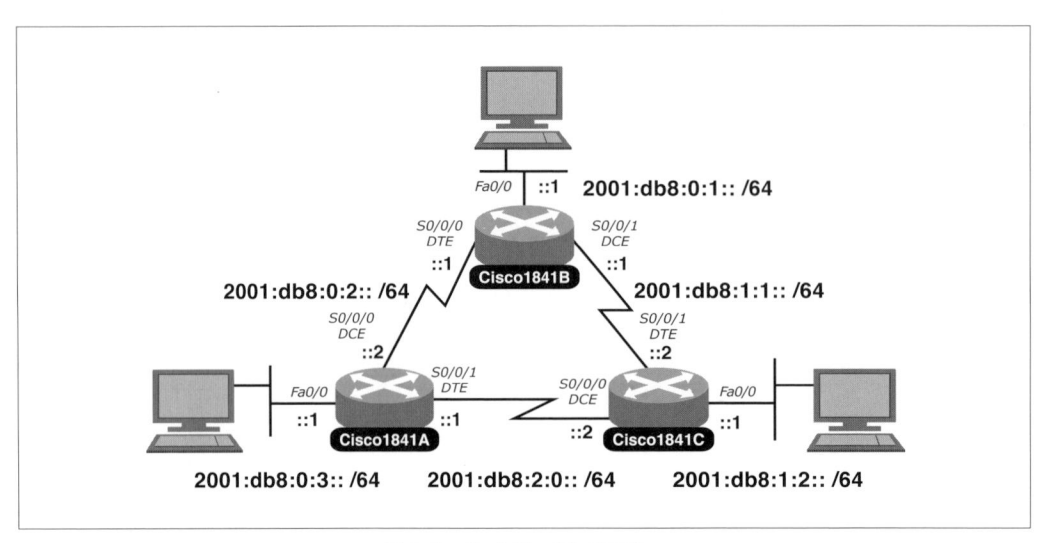

図4-2 ネットワークトポロジー

アドレスの割当は以下のとおりです。

表4-2 インターフェースアドレス割当一覧

ルーター	インターフェース	IPv6アドレス
Cisco1841A	FastEthernet 0/0 (Fa0/0)	2001:db8:0:3::1/64
	Serial 0/0/0 (S0/0/0)	2001:db8:0:2::2/64
	Serial 0/0/1 (S0/0/1)	2001:db8:2:0::1/64
Cisco1841B	FastEthernet 0/0 (Fa0/0)	2001:db8:0:1::1/64
	Serial 0/0/0 (S0/0/0)	2001:db8:0:2::1/64
	Serial 0/0/1 (S0/0/1)	2001:db8:1:1::1/64
Cisco1841C	FastEthernet 0/0 (Fa0/0)	2001:db8:1:2::1/64
	Serial 0/0/0 (S0/0/0)	2001:db8:2:0::2/64
	Serial 0/0/1 (S0/0/1)	2001:db8:1:1::2/64

STEP1 ルーティングテーブルの確認

各ルーターのルーティングテーブルを確認します。

■ Cisco1841A:

show ipv6 route コマンドで確認します。

```
Cisco1841A#show ipv6 route
IPv6 Routing Table - 8 entries
Codes: C - Connected, L - Local, S - Static, R - RIP, B - BGP
       U - Per-user Static route
       I1 - ISIS L1, I2 - ISIS L2, IA - ISIS interarea, IS - ISIS summary
```

```
        O - OSPF intra, OI - OSPF inter, OE1 - OSPF ext 1, OE2 - OSPF ext 2
        ON1 - OSPF NSSA ext 1, ON2 - OSPF NSSA ext 2
C   2001:DB8:0:2::/64 [0/0]
     via ::, Serial0/0/0
L   2001:DB8:0:2::2/128 [0/0]
     via ::, Serial0/0/0
C   2001:DB8:0:3::/64 [0/0]
     via ::, FastEthernet0/0
L   2001:DB8:0:3::1/128 [0/0]
     via ::, FastEthernet0/0
C   2001:DB8:2::/64 [0/0]
     via ::, Serial0/0/1
L   2001:DB8:2::1/128 [0/0]
     via ::, Serial0/0/1
L   FE80::/10 [0/0]
     via ::, Null0
L   FF00::/8 [0/0]
     via ::, Null0
Cisco1841A#
```

■ Cisco1841B:

show ipv6 route コマンドで確認します。

```
Cisco1841B#show ipv6 route
IPv6 Routing Table - 8 entries
Codes: C - Connected, L - Local, S - Static, R - RIP, B - BGP
       U - Per-user Static route
       I1 - ISIS L1, I2 - ISIS L2, IA - ISIS interarea, IS - ISIS summary
       O - OSPF intra, OI - OSPF inter, OE1 - OSPF ext 1, OE2 - OSPF ext 2
       ON1 - OSPF NSSA ext 1, ON2 - OSPF NSSA ext 2
C   2001:DB8:0:1::/64 [0/0]
     via ::, FastEthernet0/0
L   2001:DB8:0:1::1/128 [0/0]
     via ::, FastEthernet0/0
C   2001:DB8:0:2::/64 [0/0]
     via ::, Serial0/0/0
L   2001:DB8:0:2::1/128 [0/0]
     via ::, Serial0/0/0
C   2001:DB8:1:1::/64 [0/0]
     via ::, Serial0/0/1
L   2001:DB8:1:1::1/128 [0/0]
     via ::, Serial0/0/1
L   FE80::/10 [0/0]
     via ::, Null0
L   FF00::/8 [0/0]
     via ::, Null0
Cisco1841B#
```

■ Cisco1841C:

show ipv6 route コマンドで確認します。

```
Cisco1841C#show ipv6 route
IPv6 Routing Table - 8 entries
Codes: C - Connected, L - Local, S - Static, R - RIP, B - BGP
       U - Per-user Static route
       I1 - ISIS L1, I2 - ISIS L2, IA - ISIS interarea, IS - ISIS summary
       O - OSPF intra, OI - OSPF inter, OE1 - OSPF ext 1, OE2 - OSPF ext 2
       ON1 - OSPF NSSA ext 1, ON2 - OSPF NSSA ext 2
C   2001:DB8:1:1::/64 [0/0]
```

```
              via ::, Serial0/0/1
L    2001:DB8:1:1::2/128 [0/0]
              via ::, Serial0/0/1
C    2001:DB8:1:2::/64 [0/0]
              via ::, FastEthernet0/0
L    2001:DB8:1:2::1/128 [0/0]
              via ::, FastEthernet0/0
C    2001:DB8:2::/64 [0/0]
              via ::, Serial0/0/0
L    2001:DB8:2::2/128 [0/0]
              via ::, Serial0/0/0
L    FE80::/10 [0/0]
              via ::, Null0
L    FF00::/8 [0/0]
              via ::, Null0
Cisco1841C#
```

STEP2 スタティックルートの設定

各ルーターにスタティックルートを設定します。

■ Cisco1841A:

```
Cisco1841A(config)#ipv6 route 2001:db8:0:1::/64 2001:db8:0:2::1
Cisco1841A(config)#ipv6 route 2001:db8:1:1::/64 2001:db8:0:2::1
Cisco1841A(config)#ipv6 route 2001:db8:1:2::/64 2001:db8:2:0::2
```

■ Cisco1841B:

```
Cisco1841B(config)#ipv6 route 2001:db8:0:3::/64 2001:db8:0:2::2
Cisco1841B(config)#ipv6 route 2001:db8:2::/64 2001:db8:0:2::2
Cisco1841B(config)#ipv6 route 2001:db8:1:2::/64 2001:db8:1:1::2
```

■ Cisco1841C:

```
Cisco1841C(config)#ipv6 route 2001:db8:0:3::/64 2001:db8:2:0::1
Cisco1841C(config)#ipv6 route 2001:db8:0:2::/64 2001:db8:2:0::1
Cisco1841C(config)#ipv6 route 2001:db8:0:1::/64 2001:db8:1:1::1
```

STEP3 ルーティングテーブルの確認

各ルーターのルーティングテーブルを確認します。

■ Cisco:1841A:

`show ipv6 route` コマンドで確認します。

```
Cisco1841A#show ipv6 route
IPv6 Routing Table - 11 entries
Codes: C - Connected, L - Local, S - Static, R - RIP, B - BGP
       U - Per-user Static route
       I1 - ISIS L1, I2 - ISIS L2, IA - ISIS interarea, IS - ISIS summary
       O - OSPF intra, OI - OSPF inter, OE1 - OSPF ext 1, OE2 - OSPF ext 2
       ON1 - OSPF NSSA ext 1, ON2 - OSPF NSSA ext 2
S    2001:DB8:0:1::/64 [1/0]
              via 2001:DB8:0:2::1
C    2001:DB8:0:2::/64 [0/0]
              via ::, Serial0/0/0
L    2001:DB8:0:2::2/128 [0/0]
```

```
       via ::, Serial0/0/0
C   2001:DB8:0:3::/64 [0/0]
       via ::, FastEthernet0/0
L   2001:DB8:0:3::1/128 [0/0]
       via ::, FastEthernet0/0
S   2001:DB8:1:1::/64 [1/0]
       via 2001:DB8:0:2::1
S   2001:DB8:1:2::/64 [1/0]
       via 2001:DB8:2::2
C   2001:DB8:2::/64 [0/0]
       via ::, Serial0/0/1
L   2001:DB8:2::1/128 [0/0]
       via ::, Serial0/0/1
L   FE80::/10 [0/0]
       via ::, Null0
L   FF00::/8 [0/0]
       via ::, Null0
Cisco1841A#
```

■ Cisco1841B:

show ipv6 route コマンドで確認します。

```
Cisco1841B#show ipv6 route
IPv6 Routing Table - 11 entries
Codes: C - Connected, L - Local, S - Static, R - RIP, B - BGP
       U - Per-user Static route
       I1 - ISIS L1, I2 - ISIS L2, IA - ISIS interarea, IS - ISIS summary
       O - OSPF intra, OI - OSPF inter, OE1 - OSPF ext 1, OE2 - OSPF ext 2
       ON1 - OSPF NSSA ext 1, ON2 - OSPF NSSA ext 2
C   2001:DB8:0:1::/64 [0/0]
       via ::, FastEthernet0/0
L   2001:DB8:0:1::1/128 [0/0]
       via ::, FastEthernet0/0
C   2001:DB8:0:2::/64 [0/0]
       via ::, Serial0/0/0
L   2001:DB8:0:2::1/128 [0/0]
       via ::, Serial0/0/0
S   2001:DB8:0:3::/64 [1/0]
       via 2001:DB8:0:2::2
C   2001:DB8:1:1::/64 [0/0]
       via ::, Serial0/0/1
L   2001:DB8:1:1::1/128 [0/0]
       via ::, Serial0/0/1
S   2001:DB8:1:2::/64 [1/0]
       via 2001:DB8:1:1::2
S   2001:DB8:2::/64 [1/0]
       via 2001:DB8:0:2::2
L   FE80::/10 [0/0]
       via ::, Null0
L   FF00::/8 [0/0]
       via ::, Null0
Cisco1841B#
```

■ Cisco1841C:

show ipv6 route コマンドで確認します。

```
Cisco1841C#show ipv6 route
IPv6 Routing Table - 11 entries
Codes: C - Connected, L - Local, S - Static, R - RIP, B - BGP
```

```
           U - Per-user Static route
           I1 - ISIS L1, I2 - ISIS L2, IA - ISIS interarea, IS - ISIS summary
           O - OSPF intra, OI - OSPF inter, OE1 - OSPF ext 1, OE2 - OSPF ext 2
           ON1 - OSPF NSSA ext 1, ON2 - OSPF NSSA ext 2
S   2001:DB8:0:1::/64 [1/0]
      via 2001:DB8:1:1::1
S   2001:DB8:0:2::/64 [1/0]
      via 2001:DB8:2::1
S   2001:DB8:0:3::/64 [1/0]
      via 2001:DB8:2::1
C   2001:DB8:1:1::/64 [0/0]
      via ::, Serial0/0/1
L   2001:DB8:1:1::2/128 [0/0]
      via ::, Serial0/0/1
C   2001:DB8:1:2::/64 [0/0]
      via ::, FastEthernet0/0
L   2001:DB8:1:2::1/128 [0/0]
      via ::, FastEthernet0/0
C   2001:DB8:2::/64 [0/0]
      via ::, Serial0/0/0
L   2001:DB8:2::2/128 [0/0]
      via ::, Serial0/0/0
L   FE80::/10 [0/0]
      via ::, Null0
L   FF00::/8 [0/0]
      via ::, Null0
Cisco1841C#
```

STEP4　PC からルーターまでの通信確認

各 PC からそれぞれが接続しているルーターのインターフェースまでの通信を確認します。

■ **Cisco1841A：**

接続した PC から以下の IPv6 アドレスに対して **ping** コマンドを実行します。

```
2001:db8:0:1::1
2001:db8:1:2::1
```

■ **Cisco1841B：**

接続した PC から以下の IPv6 アドレスに対して **ping** コマンドを実行します。

```
2001:db8:0:3::1
2001:db8:1:2::1
```

■ **Cisco1841C：**

接続した PC から以下の IPv6 アドレスに対して **ping** コマンドを実行します。

```
2001:db8:0:1::1
2001:db8:0:3::1
```

Chapter 05

デフォルトルートの設定

05-1 マルチベンダー機器による実習

05-2 シスコ機器による実習

Chapter 05 デフォルトルートの設定

この章ではIPv6のデフォルトルートの設定について学びます。多くの環境においてデフォルトルートの設定は必須です。また，ダイナミックルーティングを用いる場合にも必要となります。

05-1 マルチベンダー機器による実習

第3章，第4章と同じトポロジーを用いて学習します。実習は第3章の終了時点の状態から始めます。

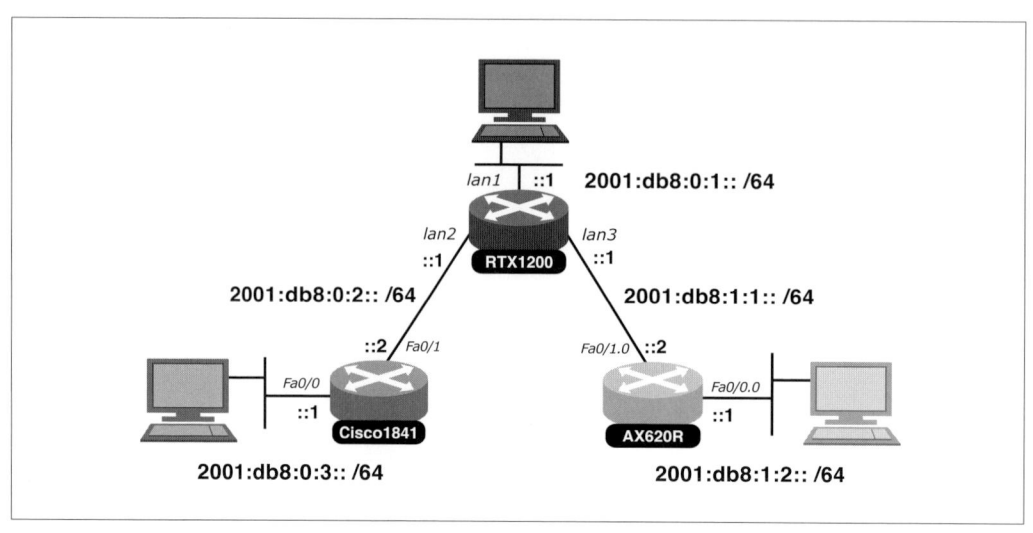

図5-1 ネットワークトポロジー

アドレスの割当は以下のとおりです。

表5-1 インターフェースアドレス割当一覧

ルーター	インターフェース	IPv6アドレス
Cisco1841	FastEthernet 0/0 (Fa0/0)	2001:db8:0:3::1/64
Cisco1841	FastEthernet 0/1 (Fa0/1)	2001:db8:0:2::2/64
RTX1200	lan1	2001:db8:0:1::1/64
RTX1200	lan2	2001:db8:0:2::1/64
RTX1200	lan3	2001:db8:1:1::1/64
AX620R	FastEthernet 0/0.0 (Fa0/0.0)	2001:db8:1:2::1/64
AX620R	FastEthernet 0/1.0 (Fa0/1.0)	2001:db8:1:1::2/64

STEP1 ルーティングテーブルの確認

各ルーターのルーティングテーブルを確認します。

■ **Cisco1841:**

`show ipv6 route` コマンドで確認します。

```
Cisco1841#show ipv6 route
IPv6 Routing Table - 6 entries
```

```
Codes: C - Connected, L - Local, S - Static, R - RIP, B - BGP
       U - Per-user Static route
       I1 - ISIS L1, I2 - ISIS L2, IA - ISIS interarea, IS - ISIS summary
       O - OSPF intra, OI - OSPF inter, OE1 - OSPF ext 1, OE2 - OSPF ext 2
       ON1 - OSPF NSSA ext 1, ON2 - OSPF NSSA ext 2
C   2001:DB8:0:2::/64 [0/0]
     via ::, FastEthernet0/1
L   2001:DB8:0:2::2/128 [0/0]
     via ::, FastEthernet0/1
C   2001:DB8:0:3::/64 [0/0]
     via ::, FastEthernet0/0
L   2001:DB8:0:3::1/128 [0/0]
     via ::, FastEthernet0/0
L   FE80::/10 [0/0]
     via ::, Null0
L   FF00::/8 [0/0]
     via ::, Null0
Cisco1841#
```

■ **RTX1200:**

show ipv6 route コマンドで確認します。

```
RTX1200#show ipv6 route
Destination            Gateway              Interface    Type
2001:db8:0:1::/64      -                    LAN1         implicit
2001:db8:0:2::/64      -                    LAN2         implicit
2001:db8:1:1::/64      -                    LAN3         implicit
RTX1200#
```

■ **AX620R:**

show ipv6 route コマンドで確認します。

```
AX620R#show ipv6 route
IPv6 Routing Table - 6 entries, unlimited
Codes: C - Connected, L - Local, S - Static
       R - RIPng, O - OSPF, IA - OSPF inter area
       E1 - OSPF external type 1, E2 - OSPF external type 2, B - BGP
       s - Summary
Timers: Uptime/Age
C     2001:db8:1:1::/64 global [0/1]
        via ::, FastEthernet0/1.0, 2:28:28/0:00:00
L     2001:db8:1:1::/128 global [0/1]
        via ::, FastEthernet0/1.0, 2:28:29/0:00:00
L     2001:db8:1:1::2/128 global [0/1]
        via ::, FastEthernet0/1.0, 2:28:28/0:00:00
C     2001:db8:1:2::/64 global [0/1]
        via ::, FastEthernet0/0.0, 0:05:49/0:00:00
L     2001:db8:1:2::/128 global [0/1]
        via ::, FastEthernet0/0.0, 0:05:50/0:00:00
L     2001:db8:1:2::1/128 global [0/1]
        via ::, FastEthernet0/0.0, 0:05:49/0:00:00
AX620R#
```

STEP2　デフォルトルートの設定

各ルーターにデフォルトルートの設定をします。

Chapter 05 デフォルトルートの設定

■ Cisco:1841:
```
Cisco1841(config)#ipv6 route ::/0 2001:db8:0:2::1
```

■ AX620R:
```
AX620R(config)#ipv6 route default 2001:db8:1:1::1
```

RTX1200 は学習のため AX620R をデフォルトルーターとして設定します。

```
RTX1200#ipv6 route default gateway 2001:db8:1:1::2%3
```

STEP3 ルーティングテーブルの確認

各ルーターのルーティングテーブルを確認します。Cisco1841 や AX620R ではデフォルトルートは "::/0" で示されます。

■ Cisco1841:

`show ipv6 route` コマンドで確認します。

```
Cisco1841#show ipv6 route
IPv6 Routing Table - 7 entries
Codes: C - Connected, L - Local, S - Static, R - RIP, B - BGP
       U - Per-user Static route
       I1 - ISIS L1, I2 - ISIS L2, IA - ISIS interarea, IS - ISIS summary
       O - OSPF intra, OI - OSPF inter, OE1 - OSPF ext 1, OE2 - OSPF ext 2
       ON1 - OSPF NSSA ext 1, ON2 - OSPF NSSA ext 2
S   ::/0 [1/0]
     via 2001:DB8:0:2::1
C   2001:DB8:0:2::/64 [0/0]
     via ::, FastEthernet0/1
L   2001:DB8:0:2::2/128 [0/0]
     via ::, FastEthernet0/1
C   2001:DB8:0:3::/64 [0/0]
     via ::, FastEthernet0/0
L   2001:DB8:0:3::1/128 [0/0]
     via ::, FastEthernet0/0
L   FE80::/10 [0/0]
     via ::, Null0
L   FF00::/8 [0/0]
     via ::, Null0
Cisco1841#
```

■ RTX1200:

`show ipv6 route` コマンドで確認します。

```
RTX1200#show ipv6 route
Destination              Gateway              Interface   Type
default                  2001:db8:1:1::2      LAN3        static
2001:db8:0:1::/64        -                    LAN1        implicit
2001:db8:0:2::/64        -                    LAN2        implicit
2001:db8:0:3::/64        2001:db8:0:2::2      LAN2        static
2001:db8:1:1::/64        -                    LAN3        implicit
RTX1200#
```

■ AX620R:

`show ipv6 route` コマンドで確認します。

```
AX620R(config)#show ipv6 route
IPv6 Routing Table - 7 entries, unlimited
Codes: C - Connected, L - Local, S - Static
       R - RIPng, O - OSPF, IA - OSPF inter area
       E1 - OSPF external type 1, E2 - OSPF external type 2, B - BGP
       s - Summary
Timers: Uptime/Age
S      ::/0 orphan [1/1]
           via 2001:db8:1:1::1, FastEthernet0/1.0, 0:10:42/0:00:00
C      2001:db8:1:1::/64 global [0/1]
           via ::, FastEthernet0/1.0, 1:16:11/0:00:00
L      2001:db8:1:1::/128 global [0/1]
           via ::, FastEthernet0/1.0, 1:16:12/0:00:00
L      2001:db8:1:1::2/128 global [0/1]
           via ::, FastEthernet0/1.0, 1:16:11/0:00:00
C      2001:db8:1:2::/64 global [0/1]
           via ::, FastEthernet0/0.0, 0:31:00/0:00:00
L      2001:db8:1:2::/128 global [0/1]
           via ::, FastEthernet0/0.0, 0:31:01/0:00:00
L      2001:db8:1:2::1/128 global [0/1]
           via ::, FastEthernet0/0.0, 0:31:00/0:00:00
AX620R(config)#
```

STEP4 PC からの通信確認

PC から直接つながっていないルーターのインターフェースまでの通信を確認します。

■ Cisco1841:

接続した PC から RTX1200，AX620R のインターフェースである以下の IPv6 アドレスに対して `ping` コマンドを実行します。

```
2001:db8:0:1::1
2001:db8:1:2::1
```

■ RTX1200:

接続した PC から Cisco1841，AX620R のインターフェースである以下の IPv6 アドレスに対して `ping` コマンドを実行します。

```
2001:db8:0:3::1
2001:db8:1:2::1
```

■ AX620R:

接続した PC から Cisco1841，RTX1200 のインターフェースである以下の IPv6 アドレスに対して `ping` コマンドを実行します。

```
2001:db8:0:1::1
2001:db8:0:3::1
```

05-2 シスコ機器による実習

シスコルーターも第3章，第4章と同じトポロジーを用いて学習します。実習は第3章の終了時点の状態から始めます。

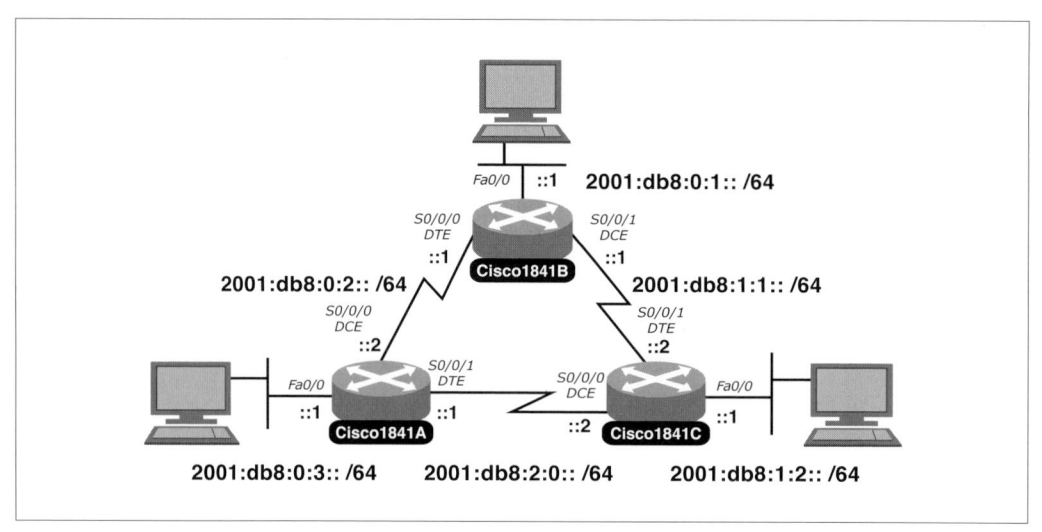

図5-2　ネットワークトポロジー

アドレスの割当は以下のとおりです。

表5-2　インターフェースアドレス割当一覧

ルーター	インターフェース	IPv6アドレス
Cisco1841A	FastEthernet 0/0 (Fa0/0)	2001:db8:0:3::1/64
	Serial 0/0/0 (S0/0/0)	2001:db8:0:2::2/64
	Serial 0/0/1 (S0/0/1)	2001:db8:2:0::1/64
Cisco1841B	FastEthernet 0/0 (Fa0/0)	2001:db8:0:1::1/64
	Serial 0/0/0 (S0/0/0)	2001:db8:0:2::1/64
	Serial 0/0/1 (S0/0/1)	2001:db8:1:1::1/64
Cisco1841C	FastEthernet 0/0 (Fa0/0)	2001:db8:1:2::1/64
	Serial 0/0/0 (S0/0/0)	2001:db8:2:0::2/64
	Serial 0/0/1 (S0/0/1)	2001:db8:1:1::2/64

STEP1　ルーティングテーブルの確認

各ルーターのルーティングテーブルを確認します。

■ Cisco1841A:

show ipv6 route コマンドで確認します。

```
Cisco1841A#show ipv6 route
IPv6 Routing Table - 8 entries
Codes: C - Connected, L - Local, S - Static, R - RIP, B - BGP
       U - Per-user Static route
       I1 - ISIS L1, I2 - ISIS L2, IA - ISIS interarea, IS - ISIS summary
```

```
            O - OSPF intra, OI - OSPF inter, OE1 - OSPF ext 1, OE2 - OSPF ext 2
            ON1 - OSPF NSSA ext 1, ON2 - OSPF NSSA ext 2
C   2001:DB8:0:2::/64 [0/0]
     via ::, Serial0/0/0
L   2001:DB8:0:2::2/128 [0/0]
     via ::, Serial0/0/0
C   2001:DB8:0:3::/64 [0/0]
     via ::, FastEthernet0/0
L   2001:DB8:0:3::1/128 [0/0]
     via ::, FastEthernet0/0
C   2001:DB8:2::/64 [0/0]
     via ::, Serial0/0/1
L   2001:DB8:2::1/128 [0/0]
     via ::, Serial0/0/1
L   FE80::/10 [0/0]
     via ::, Null0
L   FF00::/8 [0/0]
     via ::, Null0
Cisco1841A#
```

■ **Cisco1841B:**

show ipv6 route コマンドで確認します。

```
Cisco1841B#show ipv6 route
IPv6 Routing Table - 8 entries
Codes: C - Connected, L - Local, S - Static, R - RIP, B - BGP
       U - Per-user Static route
       I1 - ISIS L1, I2 - ISIS L2, IA - ISIS interarea, IS - ISIS summary
       O - OSPF intra, OI - OSPF inter, OE1 - OSPF ext 1, OE2 - OSPF ext 2
       ON1 - OSPF NSSA ext 1, ON2 - OSPF NSSA ext 2
C   2001:DB8:0:1::/64 [0/0]
     via ::, FastEthernet0/0
L   2001:DB8:0:1::1/128 [0/0]
     via ::, FastEthernet0/0
C   2001:DB8:0:2::/64 [0/0]
     via ::, Serial0/0/0
L   2001:DB8:0:2::1/128 [0/0]
     via ::, Serial0/0/0
C   2001:DB8:1:1::/64 [0/0]
     via ::, Serial0/0/1
L   2001:DB8:1:1::1/128 [0/0]
     via ::, Serial0/0/1
L   FE80::/10 [0/0]
     via ::, Null0
L   FF00::/8 [0/0]
     via ::, Null0
Cisco1841B#
```

■ **Cisco1841C:**

show ipv6 route コマンドで確認します。

```
Cisco1841C#show ipv6 route
IPv6 Routing Table - 8 entries
Codes: C - Connected, L - Local, S - Static, R - RIP, B - BGP
       U - Per-user Static route
       I1 - ISIS L1, I2 - ISIS L2, IA - ISIS interarea, IS - ISIS summary
       O - OSPF intra, OI - OSPF inter, OE1 - OSPF ext 1, OE2 - OSPF ext 2
       ON1 - OSPF NSSA ext 1, ON2 - OSPF NSSA ext 2
C   2001:DB8:1:1::/64 [0/0]
```

Chapter 05 デフォルトルートの設定

```
         via ::, Serial0/0/1
L    2001:DB8:1:1::2/128 [0/0]
         via ::, Serial0/0/1
C    2001:DB8:1:2::/64 [0/0]
         via ::, FastEthernet0/0
L    2001:DB8:1:2::1/128 [0/0]
         via ::, FastEthernet0/0
C    2001:DB8:2::/64 [0/0]
         via ::, Serial0/0/0
L    2001:DB8:2::2/128 [0/0]
         via ::, Serial0/0/0
L    FE80::/10 [0/0]
         via ::, Null0
L    FF00::/8 [0/0]
         via ::, Null0
Cisco1841C#
```

STEP2 デフォルトルートの設定

各ルーターにデフォルトルートの設定をします。ここではそれぞれ時計回り，右隣りのルーターをデフォルトルートとして設定します。

■ Cisco:1841A：

```
Cisco1841A(config)#ipv6 route ::/0 2001:db8:0:2::1
```

■ Cisco1841B：

```
Cisco1841B(config)#ipv6 route ::/0 2001:db8:1:1::2
```

■ Cisco1841C：

```
Cisco1841C(config)#ipv6 route ::/0 2001:db8:2:0::1
```

STEP3 ルーティングテーブルの確認

各ルーターのルーティングテーブルを確認します。デフォルトルートは "::/0" で示されます。

■ Cisco:1841A：

`show ipv6 route` コマンドで確認します。

```
Cisco1841A#show ipv6 route
IPv6 Routing Table - 9 entries
Codes: C - Connected, L - Local, S - Static, R - RIP, B - BGP
       U - Per-user Static route
       I1 - ISIS L1, I2 - ISIS L2, IA - ISIS interarea, IS - ISIS summary
       O - OSPF intra, OI - OSPF inter, OE1 - OSPF ext 1, OE2 - OSPF ext 2
       ON1 - OSPF NSSA ext 1, ON2 - OSPF NSSA ext 2
S    ::/0 [1/0]
         via 2001:DB8:0:2::1
C    2001:DB8:0:2::/64 [0/0]
         via ::, Serial0/0/0
L    2001:DB8:0:2::2/128 [0/0]
         via ::, Serial0/0/0
C    2001:DB8:0:3::/64 [0/0]
         via ::, FastEthernet0/0
L    2001:DB8:0:3::1/128 [0/0]
```

```
        via ::, FastEthernet0/0
C    2001:DB8:2::/64 [0/0]
        via ::, Serial0/0/1
L    2001:DB8:2::1/128 [0/0]
        via ::, Serial0/0/1
L    FE80::/10 [0/0]
        via ::, Null0
L    FF00::/8 [0/0]
        via ::, Null0
Cisco1841A#
```

■ Cisco1841B:

show ipv6 route コマンドで確認します。

```
Cisco1841B#show ipv6 route
IPv6 Routing Table - 9 entries
Codes: C - Connected, L - Local, S - Static, R - RIP, B - BGP
       U - Per-user Static route
       I1 - ISIS L1, I2 - ISIS L2, IA - ISIS interarea, IS - ISIS summary
       O - OSPF intra, OI - OSPF inter, OE1 - OSPF ext 1, OE2 - OSPF ext 2
       ON1 - OSPF NSSA ext 1, ON2 - OSPF NSSA ext 2
S    ::/0 [1/0]
        via 2001:DB8:1:1::2
C    2001:DB8:0:1::/64 [0/0]
        via ::, FastEthernet0/0
L    2001:DB8:0:1::1/128 [0/0]
        via ::, FastEthernet0/0
C    2001:DB8:0:2::/64 [0/0]
        via ::, Serial0/0/0
L    2001:DB8:0:2::1/128 [0/0]
        via ::, Serial0/0/0
C    2001:DB8:1:1::/64 [0/0]
        via ::, Serial0/0/1
L    2001:DB8:1:1::1/128 [0/0]
        via ::, Serial0/0/1
L    FE80::/10 [0/0]
        via ::, Null0
L    FF00::/8 [0/0]
        via ::, Null0
Cisco1841B#
```

■ Cisco1841C:

show ipv6 route コマンドで確認します。

```
Cisco1841C#show ipv6 route
IPv6 Routing Table - 9 entries
Codes: C - Connected, L - Local, S - Static, R - RIP, B - BGP
       U - Per-user Static route
       I1 - ISIS L1, I2 - ISIS L2, IA - ISIS interarea, IS - ISIS summary
       O - OSPF intra, OI - OSPF inter, OE1 - OSPF ext 1, OE2 - OSPF ext 2
       ON1 - OSPF NSSA ext 1, ON2 - OSPF NSSA ext 2
S    ::/0 [1/0]
        via 2001:DB8:2::1
C    2001:DB8:1:1::/64 [0/0]
        via ::, Serial0/0/1
L    2001:DB8:1:1::2/128 [0/0]
        via ::, Serial0/0/1
C    2001:DB8:1:2::/64 [0/0]
        via ::, FastEthernet0/0
```

Chapter 05 デフォルトルートの設定

```
L    2001:DB8:1:2::1/128 [0/0]
        via ::, FastEthernet0/0
C    2001:DB8:2::/64 [0/0]
        via ::, Serial0/0/0
L    2001:DB8:2::2/128 [0/0]
        via ::, Serial0/0/0
L    FE80::/10 [0/0]
        via ::, Null0
L    FF00::/8 [0/0]
        via ::, Null0
Cisco1841C#
```

STEP4　PC からの通信確認

PC から直接つながっていないルーターのインターフェースまでの通信を確認します。

■ **Cisco1841A：**

接続した PC からの Cisco1841B，Cisco1841C のインターフェースである以下の IP アドレスに対して **ping** コマンドを実行します。

```
2001:db8:0:1::1
2001:db8:1:2::1
```

■ **Cisco1841B：**

接続した PC からの Cisco1841A，Cisco1841C のインターフェースである以下の IP アドレスに対して **ping** コマンドを実行します。

```
2001:db8:0:3::1
2001:db8:1:2::1
```

■ **Cisco1841C：**

接続した PC からの Cisco1841A，Cisco1841B のインターフェースである以下の IP アドレスに対して **ping** コマンドを実行します。

```
2001:db8:0:1::1
2001:db8:0:3::1
```

Chapter 06

RIPngの設定

06-1 マルチベンダー機器
　　　による実習

06-2 シスコ機器による実習

Chapter 06 RIPngの設定

この章ではIPv6のダイナミックルーティングプロトコルであるRIPngについて学びます。RIPngはIPv4におけるRIPと同様，多くのルーターでサポートされています。

06-1 マルチベンダー機器による実習

第3章から第5章までと同じトポロジーを用いて学習します。実習は第3章の終了時点の状態から始めます。

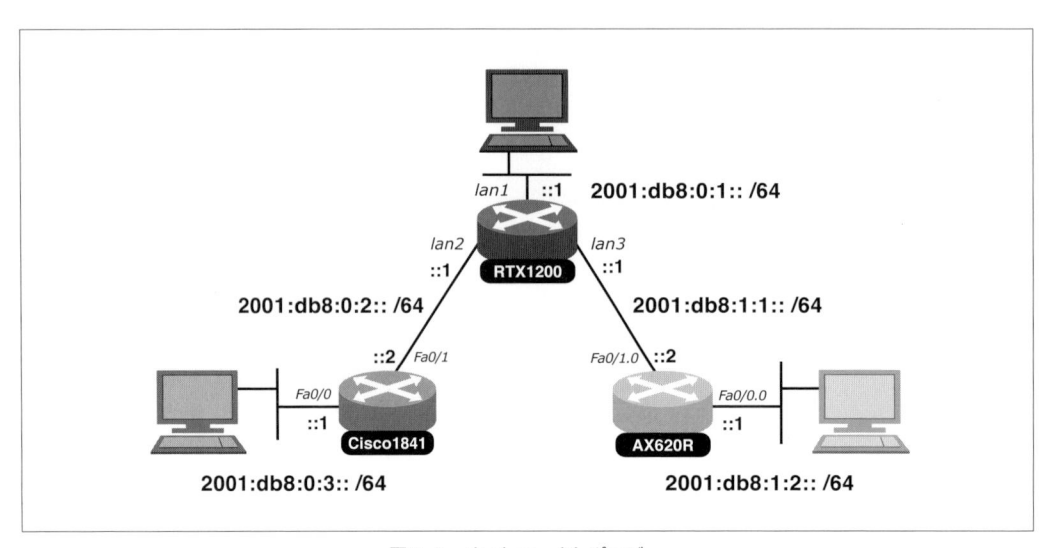

図6-1 ネットワークトポロジー

アドレスの割当は以下のとおりです。

表6-1 インターフェースアドレス割当一覧

ルーター	インターフェース	IPv6アドレス
Cisco1841	FastEthernet 0/0 (Fa0/0)	2001:db8:0:3::1/64
	FastEthernet 0/1 (Fa0/1)	2001:db8:0:2::2/64
RTX1200	lan1	2001:db8:0:1::1/64
	lan2	2001:db8:0:2::1/64
	lan3	2001:db8:1:1::1/64
AX620R	FastEthernet 0/0.0 (Fa0/0.0)	2001:db8:1:2::1/64
	FastEthernet 0/1.0 (Fa0/1.0)	2001:db8:1:1::2/64

STEP1 ルーティングテーブルの確認

各ルーターのルーティングテーブルを確認します。

■ Cisco1841:

`show ipv6 route` コマンドで確認します。

```
Cisco1841#show ipv6 route
IPv6 Routing Table - 6 entries
```

```
Codes: C - Connected, L - Local, S - Static, R - RIP, B - BGP
       U - Per-user Static route
       I1 - ISIS L1, I2 - ISIS L2, IA - ISIS interarea, IS - ISIS summary
       O - OSPF intra, OI - OSPF inter, OE1 - OSPF ext 1, OE2 - OSPF ext 2
       ON1 - OSPF NSSA ext 1, ON2 - OSPF NSSA ext 2
C   2001:DB8:0:2::/64 [0/0]
     via ::, FastEthernet0/1
L   2001:DB8:0:2::2/128 [0/0]
     via ::, FastEthernet0/1
C   2001:DB8:0:3::/64 [0/0]
     via ::, FastEthernet0/0
L   2001:DB8:0:3::1/128 [0/0]
     via ::, FastEthernet0/0
L   FE80::/10 [0/0]
     via ::, Null0
L   FF00::/8 [0/0]
     via ::, Null0
Cisco1841#
```

■ RTX1200：

show ipv6 route コマンドで確認します。

```
RTX1200#show ipv6 route
Destination               Gateway            Interface  Type
2001:db8:0:1::/64         -                  LAN1       implicit
2001:db8:0:2::/64         -                  LAN2       implicit
2001:db8:1:1::/64         -                  LAN3       implicit
RTX1200#
```

■ AX620R：

show ipv6 route コマンドで確認します。

```
AX620R#show ipv6 route
IPv6 Routing Table - 6 entries, unlimited
Codes: C - Connected, L - Local, S - Static
       R - RIPng, O - OSPF, IA - OSPF inter area
       E1 - OSPF external type 1, E2 - OSPF external type 2, B - BGP
       s - Summary
Timers: Uptime/Age
C    2001:db8:1:1::/64 global [0/1]
        via ::, FastEthernet0/1.0, 2:28:28/0:00:00
L    2001:db8:1:1::/128 global [0/1]
        via ::, FastEthernet0/1.0, 2:28:29/0:00:00
L    2001:db8:1:1::2/128 global [0/1]
        via ::, FastEthernet0/1.0, 2:28:28/0:00:00
C    2001:db8:1:2::/64 global [0/1]
        via ::, FastEthernet0/0.0, 0:05:49/0:00:00
L    2001:db8:1:2::/128 global [0/1]
        via ::, FastEthernet0/0.0, 0:05:50/0:00:00
L    2001:db8:1:2::1/128 global [0/1]
        via ::, FastEthernet0/0.0, 0:05:49/0:00:00
AX620R#
```

STEP2 RIPng プロセスの起動

IPv4 の RIP では，ルーティングプロトコルは **router rip** コマンドで起動し，ネットワークアドレスを設定することでルーティングの対象となるネットワークを指定します。これに対して IPv6 では，

Chapter 06 RIPngの設定

インターフェースの設定でルーティングの対象となるネットワークを指定します。

■ **Cisco 1841:**

```
Cisco1841(config-if)#interface fa0/1
Cisco1841(config-if)#ipv6 rip RIPng enable
```

これにより，RIPngという名前のRIPプロセスがグローバルでも有効になります。インターフェースにRIPngという名前のプロセスを割り当てる前に，名前を指定しつつRIPプロセスを明示的に起動したい場合は，次のように入力します。

```
Router(config)#ipv6 rip RIPng
```

■ **RTX1200：**

```
RTX1200#ipv6 rip use on
```

■ **AX620R：**

```
AX620R(config)#ipv6 router rip

AX620R(config)#interface fa0/0.0
AX620R(config-if)#ipv6 rip enable

AX620R(config)#interface fa0/1.0
AX620R(config-if)#ipv6 rip enable
```

STEP3　RIPngメッセージ送信の制御

いずれのルーターでも，RIPプロセスが起動されるとインターフェースに割り当てられているIPv6アドレスが経路情報としてアドバタイズされるようになります。

しかし，経路情報はアドバタイズしたいが，特定のインターフェースについては経路情報を送受したくないという場合があります。このトポロジーの例では，RTX1200のlan1，AX620Rのfa0/0.0，Cisco 1841のFa0/0などです。これらのポートにはPCしか接続しておらず，RIPのアドバタイズを送受するのは不適切だからです。

そこで，一部のポートについてRIPの経路情報を送受しない設定を行います。Ciscoルーターについては，RIPngによるpassive interfaceの設定が不可能であるため，fa0/0（PCの接続されているポート）についてRIPngを有効にしない代わりに，静的ルーティング情報を再配布することで，同じ効果を実現します。

■ **Cisco 1841：**

```
Cisco1841(config)#ipv6 router rip RIPng
Cisco1841(config-rtr)#redistribute connected
```

■ **RTX1200：**

```
RTX1200#ipv6 lan1 rip send off
RTX1200#ipv6 lan1 rip receive off
```

■ **AX620R：**

```
(config)#interface fa0/0.0
(config-if)#no ipv6 rip send
(config-if)#no ipv6 rip receive
```

STEP4 ルーティングテーブルの確認

各ルーターのルーティングテーブルを確認します。

■ Cisco 1841：

show ipv6 route コマンドで確認します。

```
Cisco1841#show ipv6 route
IPv6 Routing Table - Default - 8 entries
Codes: C - Connected, L - Local, S - Static, U - Per-user Static route
       B - BGP, M - MIPv6, R - RIP, I1 - ISIS L1
       I2 - ISIS L2, IA - ISIS interarea, IS - ISIS summary, D - EIGRP
       EX - EIGRP external
       O - OSPF Intra, OI - OSPF Inter, OE1 - OSPF ext 1, OE2 - OSPF ext 2
       ON1 - OSPF NSSA ext 1, ON2 - OSPF NSSA ext 2
R   2001:DB8:0:1::/64 [120/2]
     via FE80::2A0:DEFF:FE65:AFD2, FastEthernet0/1
C   2001:DB8:0:2::/64 [0/0]
     via FastEthernet0/1, directly connected
L   2001:DB8:0:2::2/128 [0/0]
     via FastEthernet0/1, receive
C   2001:DB8:0:3::/64 [0/0]
     via FastEthernet0/0, directly connected
L   2001:DB8:0:3::1/128 [0/0]
     via FastEthernet0/0, receive
R   2001:DB8:1:1::/64 [120/2]
     via FE80::2A0:DEFF:FE65:AFD2, FastEthernet0/1
R   2001:DB8:1:2::/64 [120/3]
     via FE80::2A0:DEFF:FE65:AFD2, FastEthernet0/1
L   FF00::/8 [0/0]
     via Null0, receive
```

■ RTX1200：

show ipv6 route コマンドで確認します。

```
RTX1200#show ipv6 route
Destination           Gateway                       Interface Type
2001:db8:0:1::/64     -                             LAN1      implicit
2001:db8:0:2::/64     -                             LAN2      implicit
2001:db8:0:3::/64     fe80::223:ebff:fe44:ed0d LAN2    RIPng
2001:db8:1:1::/64     -                             LAN3      implicit
2001:db8:1:2::/64     fe80::230:13ff:fef6:1e4e LAN3    RIPng
```

■ AX620R：

show ipv6 route コマンドで確認します。

```
AX620R(config)#show ipv6 route
IPv6 Routing Table - 9 entries, unlimited
Codes: C - Connected, L - Local, S - Static
       R - RIPng, O - OSPF, IA - OSPF inter area
       E1 - OSPF external type 1, E2 - OSPF external type 2, B - BGP
       s - Summary
Timers: Uptime/Age
R      2001:db8:0:1::/64 global [120/2]
         via fe80::2a0:deff:fe65:afd3, FastEthernet0/1.0, 0:10:35/0:00:29
R      2001:db8:0:2::/64 global [120/2]
         via fe80::2a0:deff:fe65:afd3, FastEthernet0/1.0, 0:10:35/0:00:29
R      2001:db8:0:3::/64 global [120/3]
         via fe80::2a0:deff:fe65:afd3, FastEthernet0/1.0, 0:03:45/0:00:29
```

```
C       2001:db8:1:1::/64 global [0/1]
            via ::, FastEthernet0/1.0, 0:47:32/0:00:00
L       2001:db8:1:1::/128 global [0/1]
            via ::, FastEthernet0/1.0, 0:47:33/0:00:00
L       2001:db8:1:1::2/128 global [0/1]
            via ::, FastEthernet0/1.0, 0:47:32/0:00:00
C       2001:db8:1:2::/64 global [0/1]
            via ::, FastEthernet0/0.0, 0:48:24/0:00:00
L       2001:db8:1:2::/128 global [0/1]
            via ::, FastEthernet0/0.0, 0:48:25/0:00:00
L       2001:db8:1:2::1/128 global [0/1]
            via ::, FastEthernet0/0.0, 0:48:26/0:00:00
```

STEP5　PC からの通信確認

PC から直接つながっていないルーターのインターフェースまでの通信を確認します。

■ Cisco 1841：

接続した PC から RTX1200、AX620R のインターフェースである以下の IP アドレスに対して `ping` コマンドを実行します。

```
2001:db8:0:1::1
2001:db8:1:2::1
```

■ RTX1200：

接続した PC から Cisco1841、AX620R のインターフェースである以下の IP アドレスに対して `ping` コマンドを実行します。

```
2001:db8:0:3::1
2001:db8:1:2::1
```

■ AX620R：

接続した PC から Cisco1841、RTX1200 のインターフェースである以下の IP アドレスに対して `ping` コマンドを実行します。

```
2001:db8:0:1::1
2001:db8:0:3::1
```

06-2 シスコ機器による実習

シスコルーターも第3章から第5章までと同じトポロジーを用いて学習します。実習は第3章の終了時点の状態から始めます。

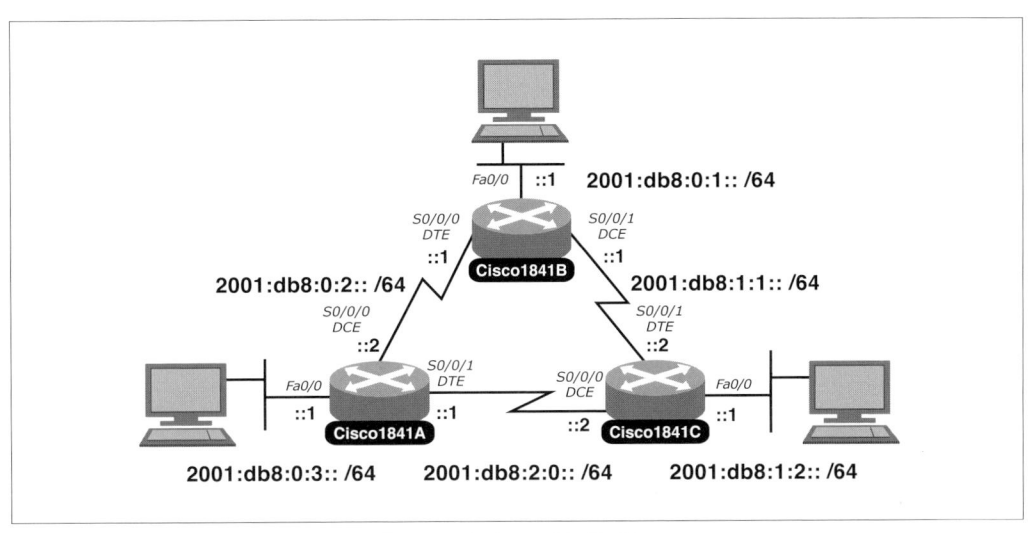

図6-2 ネットワークトポロジー

アドレスの割当は以下のとおりです。

表6-2 インターフェースアドレス割当一覧

ルーター	インターフェース	IPv6アドレス
Cisco1841A	FastEthernet 0/0 (Fa0/0)	2001:db8:0:3::1/64
	Serial 0/0/0 (S0/0/0)	2001:db8:0:2::2/64
	Serial 0/0/1 (S0/0/1)	2001:db8:2:0::1/64
Cisco1841B	FastEthernet 0/0 (Fa0/0)	2001:db8:0:1::1/64
	Serial 0/0/0 (S0/0/0)	2001:db8:0:2::1/64
	Serial 0/0/1 (S0/0/1)	2001:db8:1:1::1/64
Cisco1841C	FastEthernet 0/0 (Fa0/0)	2001:db8:1:2::1/64
	Serial 0/0/0 (S0/0/0)	2001:db8:2:0::2/64
	Serial 0/0/1 (S0/0/1)	2001:db8:1:1::2/64

STEP1 ルーティングテーブルの確認

各ルーターのルーティングテーブルを確認します。

■ Cisco1841A:

show ipv6 route コマンドで確認します。

```
Cisco1841A#show ipv6 route
IPv6 Routing Table - 8 entries
Codes: C - Connected, L - Local, S - Static, R - RIP, B - BGP
       U - Per-user Static route
       I1 - ISIS L1, I2 - ISIS L2, IA - ISIS interarea, IS - ISIS summary
```

Chapter 06 RIPngの設定

```
           O - OSPF intra, OI - OSPF inter, OE1 - OSPF ext 1, OE2 - OSPF ext 2
           ON1 - OSPF NSSA ext 1, ON2 - OSPF NSSA ext 2
C   2001:DB8:0:2::/64 [0/0]
     via ::, Serial0/0/0
L   2001:DB8:0:2::2/128 [0/0]
     via ::, Serial0/0/0
C   2001:DB8:0:3::/64 [0/0]
     via ::, FastEthernet0/0
L   2001:DB8:0:3::1/128 [0/0]
     via ::, FastEthernet0/0
C   2001:DB8:2::/64 [0/0]
     via ::, Serial0/0/1
L   2001:DB8:2::1/128 [0/0]
     via ::, Serial0/0/1
L   FE80::/10 [0/0]
     via ::, Null0
L   FF00::/8 [0/0]
     via ::, Null0
Cisco1841A#
```

■ **Cisco1841B:**

show ipv6 route コマンドで確認します。

```
Cisco1841B#show ipv6 route
IPv6 Routing Table - 8 entries
Codes: C - Connected, L - Local, S - Static, R - RIP, B - BGP
       U - Per-user Static route
       I1 - ISIS L1, I2 - ISIS L2, IA - ISIS interarea, IS - ISIS summary
       O - OSPF intra, OI - OSPF inter, OE1 - OSPF ext 1, OE2 - OSPF ext 2
       ON1 - OSPF NSSA ext 1, ON2 - OSPF NSSA ext 2
C   2001:DB8:0:1::/64 [0/0]
     via ::, FastEthernet0/0
L   2001:DB8:0:1::1/128 [0/0]
     via ::, FastEthernet0/0
C   2001:DB8:0:2::/64 [0/0]
     via ::, Serial0/0/0
L   2001:DB8:0:2::1/128 [0/0]
     via ::, Serial0/0/0
C   2001:DB8:1:1::/64 [0/0]
     via ::, Serial0/0/1
L   2001:DB8:1:1::1/128 [0/0]
     via ::, Serial0/0/1
L   FE80::/10 [0/0]
     via ::, Null0
L   FF00::/8 [0/0]
     via ::, Null0
Cisco1841B#
```

■ **Cisco1841C:**

show ipv6 route コマンドで確認します。

```
Cisco1841C#show ipv6 route
IPv6 Routing Table - 8 entries
Codes: C - Connected, L - Local, S - Static, R - RIP, B - BGP
       U - Per-user Static route
       I1 - ISIS L1, I2 - ISIS L2, IA - ISIS interarea, IS - ISIS summary
       O - OSPF intra, OI - OSPF inter, OE1 - OSPF ext 1, OE2 - OSPF ext 2
       ON1 - OSPF NSSA ext 1, ON2 - OSPF NSSA ext 2
C   2001:DB8:1:1::/64 [0/0]
```

```
         via ::, Serial0/0/1
L    2001:DB8:1:1::2/128 [0/0]
         via ::, Serial0/0/1
C    2001:DB8:1:2::/64 [0/0]
         via ::, FastEthernet0/0
L    2001:DB8:1:2::1/128 [0/0]
         via ::, FastEthernet0/0
C    2001:DB8:2::/64 [0/0]
         via ::, Serial0/0/0
L    2001:DB8:2::2/128 [0/0]
         via ::, Serial0/0/0
L    FE80::/10 [0/0]
         via ::, Null0
L    FF00::/8 [0/0]
         via ::, Null0
Cisco1841C#
```

STEP2 RIPng プロセスの起動

IPv4 の RIP では，ルーティングプロトコルは **router rip** コマンドで起動し，ネットワークアドレスを設定することでルーティングの対象となるネットワークを指定します。これに対して IPv6 では，インターフェースの設定でルーティングの対象となるネットワークを指定します。以下は Cisco1841A の設定ですが，Cisco1841B, Cisco1841C についても同様です。

■ Cisco 1841A：

```
Cisco1841A(config-if)#interface serial 0/0/0
Cisco1841A(config-if)#ipv6 rip RIPng enable
Cisco1841A(config-if)#interface serial 0/0/1
Cisco1841A(config-if)#ipv6 rip RIPng enable
```

これにより，RIPng という名前の RIP プロセスがグローバルでも有効になります。インターフェースに RIPng という名前のプロセスを割り当てる前に，名前を指定しつつ RIP プロセスを明示的に起動したい場合は，次のように入力します。

```
Cisco1841A(config)#ipv6 router rip RIPng
```

STEP3 RIPng メッセージ送信の制御

いずれのルーターでも，RIP プロセスが起動されるとインターフェースに割り当てられている IPv6 アドレスが経路情報としてアドバタイズされるようになります。

しかし，経路情報はアドバタイズしたいが，特定のインターフェースについては経路情報を送受したくないという場合があります。

この場合 IPv4 の RIP と異なり passive interface の設定がありませんので，fa0/0 においては **ipv6 rip RIPng enable** コマンドを設定せず，ルーター設定モードの中で **redistribute connected** コマンドを設定することで同様の動きを実現します。以下の例は Cisco1841A の例ですが，Cisco1841B, Cisco1841C についても同様です。

■ Cisco 1841A：

```
Cisco1841A(config)#ipv6 router rip RIPng
Cisco1841A(config-rtr)#redistribute connected
```

Chapter 06 RIPngの設定

STEP4 ルーティングテーブルの確認

各ルーターのルーティングテーブルを確認します。

■ **Cisco 1841A：**

`show ipv6 route` コマンドで確認します。

```
Cisco1841A#show ipv6 route
IPv6 Routing Table - 11 entries
Codes: C - Connected, L - Local, S - Static, R - RIP, B - BGP
       U - Per-user Static route
       I1 - ISIS L1, I2 - ISIS L2, IA - ISIS interarea, IS - ISIS summary
       O - OSPF intra, OI - OSPF inter, OE1 - OSPF ext 1, OE2 - OSPF ext 2
       ON1 - OSPF NSSA ext 1, ON2 - OSPF NSSA ext 2
R   2001:DB8:0:1::/64 [120/2]
     via FE80::225:45FF:FEF7:BC80, Serial0/0/0
C   2001:DB8:0:2::/64 [0/0]
     via ::, Serial0/0/0
L   2001:DB8:0:2::2/128 [0/0]
     via ::, Serial0/0/0
C   2001:DB8:0:3::/64 [0/0]
     via ::, FastEthernet0/0
L   2001:DB8:0:3::1/128 [0/0]
     via ::, FastEthernet0/0
R   2001:DB8:1:1::/64 [120/2]
     via FE80::225:45FF:FEF7:BC80, Serial0/0/0
R   2001:DB8:1:2::/64 [120/3]
     via FE80::225:45FF:FEF7:BC80, Serial0/0/0
C   2001:DB8:2::/64 [0/0]
     via ::, Serial0/0/1
L   2001:DB8:2::1/128 [0/0]
     via ::, Serial0/0/1
L   FE80::/10 [0/0]
     via ::, Null0
L   FF00::/8 [0/0]
     via ::, Null0
Cisco1841A#
```

■ **Cisco1841B：**

`show ipv6 route` コマンドで確認します。

```
Cisco1841B#show ipv6 route
IPv6 Routing Table - 11 entries
Codes: C - Connected, L - Local, S - Static, R - RIP, B - BGP
       U - Per-user Static route
       I1 - ISIS L1, I2 - ISIS L2, IA - ISIS interarea, IS - ISIS summary
       O - OSPF intra, OI - OSPF inter, OE1 - OSPF ext 1, OE2 - OSPF ext 2
       ON1 - OSPF NSSA ext 1, ON2 - OSPF NSSA ext 2
C   2001:DB8:0:1::/64 [0/0]
     via ::, FastEthernet0/0
L   2001:DB8:0:1::1/128 [0/0]
     via ::, FastEthernet0/0
C   2001:DB8:0:2::/64 [0/0]
     via ::, Serial0/0/0
L   2001:DB8:0:2::1/128 [0/0]
     via ::, Serial0/0/0
R   2001:DB8:0:3::/64 [120/2]
     via FE80::223:EBFF:FE44:ED0C, Serial0/0/0
```

```
C   2001:DB8:1:1::/64 [0/0]
      via ::, Serial0/0/1
L   2001:DB8:1:1::1/128 [0/0]
      via ::, Serial0/0/1
R   2001:DB8:1:2::/64 [120/2]
      via FE80::222:55FF:FE52:B7AE, Serial0/0/1
R   2001:DB8:2::/64 [120/2]
      via FE80::223:EBFF:FE44:ED0C, Serial0/0/0
      via FE80::222:55FF:FE52:B7AE, Serial0/0/1
L   FE80::/10 [0/0]
      via ::, Null0
L   FF00::/8 [0/0]
      via ::, Null0
Cisco1841B#
```

■ Cisco1841C：

show ipv6 route コマンドで確認します。

```
Cisco1841C#show ipv6 route
IPv6 Routing Table - 11 entries
Codes: C - Connected, L - Local, S - Static, R - RIP, B - BGP
       U - Per-user Static route
       I1 - ISIS L1, I2 - ISIS L2, IA - ISIS interarea, IS - ISIS summary
       O - OSPF intra, OI - OSPF inter, OE1 - OSPF ext 1, OE2 - OSPF ext 2
       ON1 - OSPF NSSA ext 1, ON2 - OSPF NSSA ext 2
R   2001:DB8:0:1::/64 [120/2]
      via FE80::222:55FF:FE52:B23C, Serial0/0/1
R   2001:DB8:0:2::/64 [120/2]
      via FE80::223:EBFF:FE44:ED0C, Serial0/0/0
      via FE80::222:55FF:FE52:B23C, Serial0/0/1
R   2001:DB8:0:3::/64 [120/2]
      via FE80::223:EBFF:FE44:ED0C, Serial0/0/0
C   2001:DB8:1:1::/64 [0/0]
      via ::, Serial0/0/1
L   2001:DB8:1:1::2/128 [0/0]
      via ::, Serial0/0/1
C   2001:DB8:1:2::/64 [0/0]
      via ::, FastEthernet0/0
L   2001:DB8:1:2::1/128 [0/0]
      via ::, FastEthernet0/0
C   2001:DB8:2::/64 [0/0]
      via ::, Serial0/0/0
L   2001:DB8:2::2/128 [0/0]
      via ::, Serial0/0/0
L   FE80::/10 [0/0]
      via ::, Null0
L   FF00::/8 [0/0]
      via ::, Null0
Cisco1841C#
```

STEP5 PCからの通信確認

PCから直接つながっていないルーターのインターフェースまでの通信を確認します。

■ Cisco 1841A：

接続したPCからCisco1841B，Cisco1841Cのインターフェースである以下のIPアドレスに対して ping コマンドを実行します。

```
2001:db8:0:1::1
2001:db8:1:2::1
```

■ **Cisco1841B：**

接続したPCからCisco1841A，Cisco1841Cのインターフェースである以下のIPアドレスに対して **ping** コマンドを実行します。

```
2001:db8:0:3::1
2001:db8:1:2::1
```

■ **Cisco1841C：**

接続したPCからCisco1841A，Cisco1841Bのインターフェースである以下のIPアドレスに対して **ping** コマンドを実行します。

```
2001:db8:0:1::1
2001:db8:0:3::1
```

Chapter 07

RIPngの
デフォルトルートの伝搬

07-1 マルチベンダー機器
　　 による実習

07-2 シスコ機器による実習

Chapter 07 RIPngのデフォルトルートの伝搬

この章では各ルーターにおけるデフォルトルートの設定をRIPngに伝搬する方法について学びます。

07-1 マルチベンダー機器による実習

ここでは第6章と同じトポロジーを用いて学習します。実習は第6章の終了時点の状態から始めます。

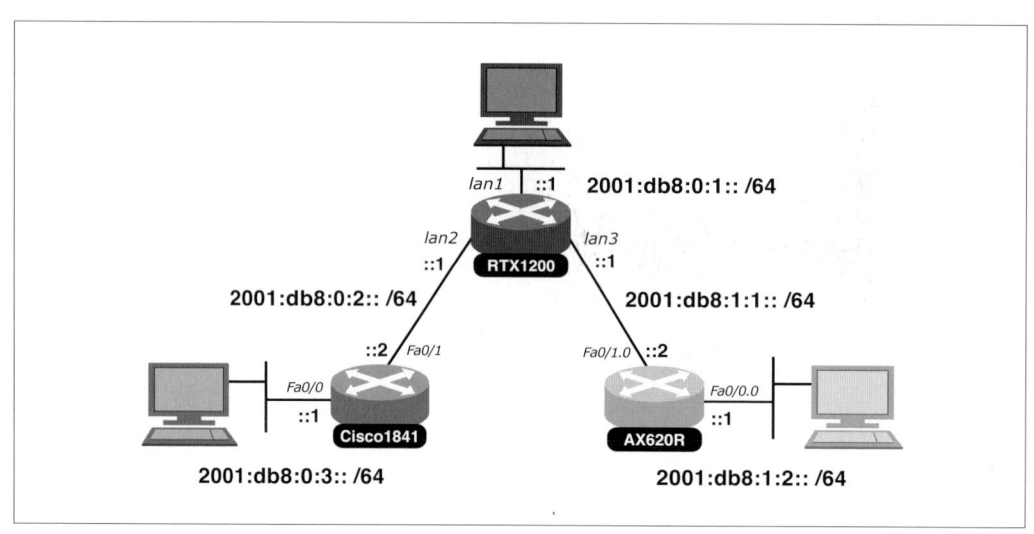

図5-1 ネットワークトポロジー

アドレスの割当は以下のとおりです。

表5-1 インターフェースアドレス割当一覧

ルーター	インターフェース	IPv6アドレス
Cisco1841	FastEthernet 0/0 (Fa0/0)	2001:db8:0:3::1/64
	FastEthernet 0/1 (Fa0/1)	2001:db8:0:2::2/64
RTX1200	lan1	2001:db8:0:1::1/64
	lan2	2001:db8:0:2::1/64
	lan3	2001:db8:1:1::1/64
AX620R	FastEthernet 0/0.0 (Fa0/0.0)	2001:db8:1:2::1/64
	FastEthernet 0/1.0 (Fa0/1.0)	2001:db8:1:1::2/64

STEP1 ルーティングテーブルの確認

各ルーターのルーティングテーブルを確認します。

■ Cisco 1841：

`show ipv6 route` コマンドで確認します。

```
Cisco1841#show ipv6 route
IPv6 Routing Table - Default - 8 entries
Codes: C - Connected, L - Local, S - Static, U - Per-user Static route
```

```
           B - BGP, M - MIPv6, R - RIP, I1 - ISIS L1
           I2 - ISIS L2, IA - ISIS interarea, IS - ISIS summary, D - EIGRP
           EX - EIGRP external
           O - OSPF Intra, OI - OSPF Inter, OE1 - OSPF ext 1, OE2 - OSPF ext 2
           ON1 - OSPF NSSA ext 1, ON2 - OSPF NSSA ext 2
R    2001:DB8:0:1::/64 [120/2]
       via FE80::2A0:DEFF:FE65:AFD2, FastEthernet0/1
C    2001:DB8:0:2::/64 [0/0]
       via FastEthernet0/1, directly connected
L    2001:DB8:0:2::2/128 [0/0]
       via FastEthernet0/1, receive
C    2001:DB8:0:3::/64 [0/0]
       via FastEthernet0/0, directly connected
L    2001:DB8:0:3::1/128 [0/0]
       via FastEthernet0/0, receive
R    2001:DB8:1:1::/64 [120/2]
       via FE80::2A0:DEFF:FE65:AFD2, FastEthernet0/1
R    2001:DB8:1:2::/64 [120/3]
       via FE80::2A0:DEFF:FE65:AFD2, FastEthernet0/1
L    FF00::/8 [0/0]
       via Null0, receive
```

■RTX1200：

show ipv6 route コマンドで確認します。

```
RTX1200#show ipv6 route
Destination              Gateway                        Interface   Type
2001:db8:0:1::/64        -                              LAN1        implicit
2001:db8:0:2::/64        -                              LAN2        implicit
2001:db8:0:3::/64        fe80::223:ebff:fe44:ed0d LAN2  RIPng
2001:db8:1:1::/64        -                              LAN3        implicit
2001:db8:1:2::/64        fe80::230:13ff:fef6:1e4e LAN3  RIPng
```

■AX620R：

show ipv6 route コマンドで確認します。

```
AX620R(config)#show ipv6 route
IPv6 Routing Table - 9 entries, unlimited
Codes: C - Connected, L - Local, S - Static
       R - RIPng, O - OSPF, IA - OSPF inter area
       E1 - OSPF external type 1, E2 - OSPF external type 2, B - BGP
       s - Summary
Timers: Uptime/Age
R     2001:db8:0:1::/64 global [120/2]
         via fe80::2a0:deff:fe65:afd3, FastEthernet0/1.0, 0:10:35/0:00:29
R     2001:db8:0:2::/64 global [120/2]
         via fe80::2a0:deff:fe65:afd3, FastEthernet0/1.0, 0:10:35/0:00:29
R     2001:db8:0:3::/64 global [120/3]
         via fe80::2a0:deff:fe65:afd3, FastEthernet0/1.0, 0:03:45/0:00:29
C     2001:db8:1:1::/64 global [0/1]
         via ::, FastEthernet0/1.0, 0:47:32/0:00:00
L     2001:db8:1:1::/128 global [0/1]
         via ::, FastEthernet0/1.0, 0:47:33/0:00:00
L     2001:db8:1:1::2/128 global [0/1]
         via ::, FastEthernet0/1.0, 0:47:32/0:00:00
C     2001:db8:1:2::/64 global [0/1]
         via ::, FastEthernet0/0.0, 0:48:24/0:00:00
L     2001:db8:1:2::/128 global [0/1]
         via ::, FastEthernet0/0.0, 0:48:25/0:00:00
```

Chapter 07 RIPngのデフォルトルートの伝搬

```
L    2001:db8:1:2::1/128 global [0/1]
       via ::, FastEthernet0/0.0, 0:48:26/0:00:00
```

STEP2 Cisco1841によるデフォルトルートの伝搬

まずCisco1841のFa0/0をデフォルトルートとみなし，その他のルーターにその情報を流す実習を行います。そのために静的にfa0/0の経路をデフォルトルートと設定しその後，その静的設定をRIPngに再配布する方法を取ります。さらにfastethernet 0/0 インターフェースのRIPを停止し，fastethernet 0/1側のRIPngでデフォルトルート情報をアドバタイズさせます。

■ Cisco 1841：

```
Cisco1841(config)#ipv6 route ::/0 fastethernet 0/0
Cisco1841(config)#ipv6 router rip RIPng
Cisco1841(config-rtr)#redistribute static
Cisco1841(config)#interface fastethernet 0/0
Cisco1841(config-if)#no ipv6 rip RIPng enable
Cisco1841(config-if)#interface fastethernet 0/0
Cisco1841(config-if)#ipv6 rip RIPng default-information originate
```

なお，他のルーターはすでにRIPngが動作しているため，追加の設定は不要です。設定が終了したら，各ルーターのルーティングテーブルにデフォルトルートの存在を確認します。

■ Cisco1841：

show ipv6 route コマンドで確認します。

```
Cisco1841#show ipv6 route
IPv6 Routing Table - Default - 9 entries
Codes: C - Connected, L - Local, S - Static, U - Per-user Static route
       B - BGP, M - MIPv6, R - RIP, I1 - ISIS L1
       I2 - ISIS L2, IA - ISIS interarea, IS - ISIS summary, D - EIGRP
       EX - EIGRP external
       O - OSPF Intra, OI - OSPF Inter, OE1 - OSPF ext 1, OE2 - OSPF ext 2
       ON1 - OSPF NSSA ext 1, ON2 - OSPF NSSA ext 2
S   ::/0 [1/0]
     via FastEthernet0/0, directly connected
R   2001:DB8:0:1::/64 [120/2]
     via FE80::2A0:DEFF:FE37:80DC, FastEthernet0/1
C   2001:DB8:0:2::/64 [0/0]
     via FastEthernet0/1, directly connected
L   2001:DB8:0:2::2/128 [0/0]
     via FastEthernet0/1, receive
C   2001:DB8:0:3::/64 [0/0]
     via FastEthernet0/0, directly connected
L   2001:DB8:0:3::1/128 [0/0]
     via FastEthernet0/0, receive
R   2001:DB8:1:1::/64 [120/2]
     via FE80::2A0:DEFF:FE37:80DC, FastEthernet0/1
R   2001:DB8:1:2::/64 [120/3]
     via FE80::2A0:DEFF:FE37:80DC, FastEthernet0/1
L   FF00::/8 [0/0]
     via Null0, receive
```

■ RTX1200：

show ipv6 route コマンドで確認します。

```
RTX1200#show ipv6 route
Destination              Gateway                        Interface   Type
default                  fe80::46e4:d9ff:fe17:ff77      LAN2        RIPng
2001:db8:0:1::/64        -                              LAN1        implicit
2001:db8:0:2::/64        -                              LAN2        implicit
2001:db8:1:1::/64        -                              LAN3        implicit
2001:db8:1:2::/64        fe80::260:b9ff:fe4f:22df       LAN3        RIPng
```

■ AX620R：

show ipv6 route コマンドで確認します。

```
AX620R(config)#show ipv6 route
IPv6 Routing Table - 9 entries, unlimited
Codes: C - Connected, L - Local, S - Static
       R - RIPng, O - OSPF, IA - OSPF inter area
       E1 - OSPF external type 1, E2 - OSPF external type 2, B - BGP
       s - Summary
Timers: Uptime/Age
R       ::/0 orphan [120/3]
          via fe80::2a0:deff:fe37:80dd, FastEthernet0/1.0, 0:00:28/0:00:28
R       2001:db8:0:1::/64 global [120/2]
          via fe80::2a0:deff:fe37:80dd, FastEthernet0/1.0, 0:00:28/0:00:28
R       2001:db8:0:2::/64 global [120/2]
          via fe80::2a0:deff:fe37:80dd, FastEthernet0/1.0, 0:00:28/0:00:28
C       2001:db8:1:1::/64 global [0/1]
          via ::, FastEthernet0/1.0, 1d48m39s/0:00:00
L       2001:db8:1:1::/128 global [0/1]
          via ::, FastEthernet0/1.0, 1d48m40s/0:00:00
L       2001:db8:1:1::2/128 global [0/1]
          via ::, FastEthernet0/1.0, 1d48m39s/0:00:00
C       2001:db8:1:2::/64 global [0/1]
          via ::, FastEthernet0/0.0, 1d48m30s/0:00:00
L       2001:db8:1:2::/128 global [0/1]
          via ::, FastEthernet0/0.0, 1d48m31s/0:00:00
L       2001:db8:1:2::1/128 global [0/1]
          via ::, FastEthernet0/0.0, 1d48m32s/0:00:00
```

STEP3 RTX1200によるデフォルトルートの伝搬

RTX1200のLan1をデフォルトルートに設定します。最初にSTEP2におけるCisco1841の設定情報を消去します。その後，下記コマンドを実行し，デフォルトルート情報をアドバタイズします。

■ Cisco 1841：

```
Cisco1841(config)#no ipv6 route ::/0 fastethernet 0/0
```

■ RTX1200：

```
RTX1200#ipv6 route default gateway 2001:db8:0:1::1%1
```

なお，他のルーターはすでにRIPが動作しているため，追加の設定は不要です。設定が終了したら，各ルーターのルーティングテーブルにデフォルトルートの存在を確認します。

Chapter 07 RIPngのデフォルトルートの伝搬

■ Cisco1841：

show ipv6 route コマンドで確認します。

```
Cisco1841#show ipv6 route
IPv6 Routing Table - Default - 9 entries
Codes: C - Connected, L - Local, S - Static, U - Per-user Static route
       B - BGP, M - MIPv6, R - RIP, I1 - ISIS L1
       I2 - ISIS L2, IA - ISIS interarea, IS - ISIS summary, D - EIGRP
       EX - EIGRP external
       O - OSPF Intra, OI - OSPF Inter, OE1 - OSPF ext 1, OE2 - OSPF ext 2
       ON1 - OSPF NSSA ext 1, ON2 - OSPF NSSA ext 2
R   ::/0 [120/2]
     via FE80::2A0:DEFF:FE37:80DC, FastEthernet0/1
R   2001:DB8:0:1::/64 [120/2]
     via FE80::2A0:DEFF:FE37:80DC, FastEthernet0/1
C   2001:DB8:0:2::/64 [0/0]
     via FastEthernet0/1, directly connected
L   2001:DB8:0:2::2/128 [0/0]
     via FastEthernet0/1, receive
C   2001:DB8:0:3::/64 [0/0]
     via FastEthernet0/0, directly connected
L   2001:DB8:0:3::1/128 [0/0]
     via FastEthernet0/0, receive
R   2001:DB8:1:1::/64 [120/2]
     via FE80::2A0:DEFF:FE37:80DC, FastEthernet0/1
R   2001:DB8:1:2::/64 [120/3]
     via FE80::2A0:DEFF:FE37:80DC, FastEthernet0/1
L   FF00::/8 [0/0]
     via Null0, receive
```

■ RTX1200：

show ipv6 route コマンドで確認します。

```
RTX1200#show ipv6 route
Destination             Gateway                       Interface  Type
default                 2001:db8:0:1::1               LAN1       static
2001:db8:0:1::/64       -                             LAN1       implicit
2001:db8:0:2::/64       -                             LAN2       implicit
2001:db8:0:3::/64       fe80::46e4:d9ff:fe17:ff77
                                                      LAN2       RIPng
2001:db8:1:1::/64       -                             LAN3       implicit
2001:db8:1:2::/64       fe80::260:b9ff:fe4f:22df LAN3       RIPng
```

■ AX620R：

show ipv6 route コマンドで確認します。

```
AX620R(config)#show ipv6 route
IPv6 Routing Table - 10 entries, unlimited
Codes: C - Connected, L - Local, S - Static
       R - RIPng, O - OSPF, IA - OSPF inter area
       E1 - OSPF external type 1, E2 - OSPF external type 2, B - BGP
       s - Summary
Timers: Uptime/Age
R     ::/0 orphan [120/2]
        via fe80::2a0:deff:fe37:80dd, FastEthernet0/1.0, 0:00:05/0:00:05
R     2001:db8:0:1::/64 global [120/2]
        via fe80::2a0:deff:fe37:80dd, FastEthernet0/1.0, 0:00:05/0:00:05
R     2001:db8:0:2::/64 global [120/2]
        via fe80::2a0:deff:fe37:80dd, FastEthernet0/1.0, 0:00:05/0:00:05
```

```
R     2001:db8:0:3::/64 global [120/3]
        via fe80::2a0:deff:fe37:80dd, FastEthernet0/1.0, 0:00:05/0:00:05
C     2001:db8:1:1::/64 global [0/1]
        via ::, FastEthernet0/1.0, 1d57m4s/0:00:00
L     2001:db8:1:1::/128 global [0/1]
        via ::, FastEthernet0/1.0, 1d57m5s/0:00:00
L     2001:db8:1:1::2/128 global [0/1]
        via ::, FastEthernet0/1.0, 1d57m4s/0:00:00
C     2001:db8:1:2::/64 global [0/1]
        via ::, FastEthernet0/0.0, 1d56m55s/0:00:00
L     2001:db8:1:2::/128 global [0/1]
        via ::, FastEthernet0/0.0, 1d56m58s/0:00:00
L     2001:db8:1:2::1/128 global [0/1]
        via ::, FastEthernet0/0.0, 1d56m57s/0:00:00
```

STEP4 AX620R によるデフォルトルートの伝搬

AX620R の FastEthernet 0/0.0 をデフォルトルートとします。STEP3 で RTX1200 に設定したデフォルトルートを消去してから以下を実行してください。

■ RTX1200：

```
RTX1200#no ipv6 route default gateway 2001:db8:0:1::1%1
```

■ AX620R：

```
AX620R(config)#ipv6 route ::/0 FastEthernet0/0.0
```

また FastEtheret 0/0.0 の RIP を停止します。

```
AX620R(config)#interface FastEthernet0/0.0
AX620R(config-FastEthernet0/0.0)#no ipv6 rip enable
```

デフォルトルートを RIPng 側にアドバタイズさせます。

```
AX620R(config)#ipv6 router rip
AX620R(config-ipv6-rip)#originate-default
```

なお，他のルーターはすでに RIPng が動作しているため，追加の設定は不要です。設定が終了したら，各ルーターのルーティングテーブルにデフォルトルートの存在を確認します。

■ Cisco1841：

`show ipv6 route` コマンドで確認します。

```
Cisco1841#show ipv6 route
IPv6 Routing Table - Default - 8 entries
Codes: C - Connected, L - Local, S - Static, U - Per-user Static route
       B - BGP, M - MIPv6, R - RIP, I1 - ISIS L1
       I2 - ISIS L2, IA - ISIS interarea, IS - ISIS summary, D - EIGRP
       EX - EIGRP external
       O - OSPF Intra, OI - OSPF Inter, OE1 - OSPF ext 1, OE2 - OSPF ext 2
       ON1 - OSPF NSSA ext 1, ON2 - OSPF NSSA ext 2
R   ::/0 [120/3]
     via FE80::2A0:DEFF:FE37:80DC, FastEthernet0/1
R   2001:DB8:0:1::/64 [120/2]
     via FE80::2A0:DEFF:FE37:80DC, FastEthernet0/1
C   2001:DB8:0:2::/64 [0/0]
     via FastEthernet0/1, directly connected
L   2001:DB8:0:2::2/128 [0/0]
```

```
            via FastEthernet0/1, receive
C    2001:DB8:0:3::/64 [0/0]
            via FastEthernet0/0, directly connected
L    2001:DB8:0:3::1/128 [0/0]
            via FastEthernet0/0, receive
R    2001:DB8:1:1::/64 [120/2]
            via FE80::2A0:DEFF:FE37:80DC, FastEthernet0/1
L    FF00::/8 [0/0]
            via Null0, receive
```

■ RTX1200：

show ipv6 route コマンドで確認します。

```
RTX1200#show ipv6 route
Destination              Gateway                      Interface  Type
default                  fe80::260:b9ff:fe4f:22df     LAN3       RIPng
2001:db8:0:1::/64        -                            LAN1       implicit
2001:db8:0:2::/64        -                            LAN2       implicit
2001:db8:0:3::/64        fe80::46e4:d9ff:fe17:ff77
                                                      LAN2       RIPng
2001:db8:1:1::/64        -                            LAN3       implicit
```

■ AX620R：

show ipv6 route コマンドで確認します。

```
AX620R(config)#show ipv6 route
IPv6 Routing Table - 10 entries, unlimited
Codes: C - Connected, L - Local, S - Static
       R - RIPng, O - OSPF, IA - OSPF inter area
       E1 - OSPF external type 1, E2 - OSPF external type 2, B - BGP
       s - Summary
Timers: Uptime/Age
S     ::/0 orphan [1/1]
          via ::, FastEthernet0/0.0, 0:03:39/0:00:00
R     2001:db8:0:1::/64 global [120/2]
          via fe80::2a0:deff:fe37:80dd, FastEthernet0/1.0, 0:08:22/0:00:23
R     2001:db8:0:2::/64 global [120/2]
          via fe80::2a0:deff:fe37:80dd, FastEthernet0/1.0, 0:08:22/0:00:23
R     2001:db8:0:3::/64 global [120/3]
          via fe80::2a0:deff:fe37:80dd, FastEthernet0/1.0, 0:08:23/0:00:24
C     2001:db8:1:1::/64 global [0/1]
          via ::, FastEthernet0/1.0, 1d1h5m21s/0:00:00
L     2001:db8:1:1::/128 global [0/1]
          via ::, FastEthernet0/1.0, 1d1h5m22s/0:00:00
L     2001:db8:1:1::2/128 global [0/1]
          via ::, FastEthernet0/1.0, 1d1h5m21s/0:00:00
C     2001:db8:1:2::/64 global [0/1]
          via ::, FastEthernet0/0.0, 1d1h5m12s/0:00:00
L     2001:db8:1:2::/128 global [0/1]
          via ::, FastEthernet0/0.0, 1d1h5m15s/0:00:00
L     2001:db8:1:2::1/128 global [0/1]
          via ::, FastEthernet0/0.0, 1d1h5m14s/0:00:00
```

07-2 シスコ機器による実習

シスコルーターも第6章と同じトポロジーを用いて学習します。実習は第6章の終了時点の状態から始めます。

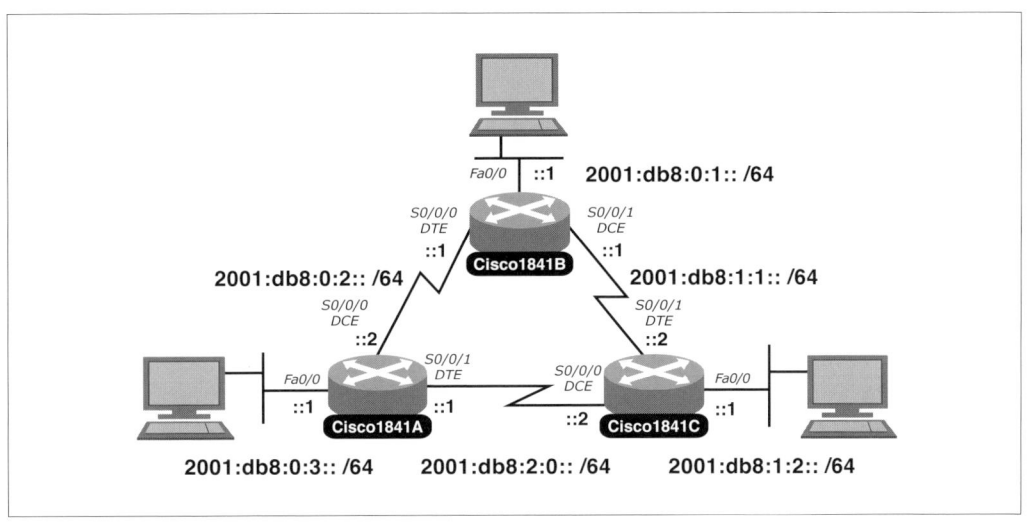

図7-2　ネットワークトポロジー

アドレスの割当は以下のとおりです。

表7-2　インターフェースアドレス割当一覧

ルーター	インターフェース	IPv6アドレス
Cisco1841A	FastEthernet 0/0 (Fa0/0)	2001:db8:0:3::1/64
	Serial 0/0/0 (S0/0/0)	2001:db8:0:2::2/64
	Serial 0/0/1 (S0/0/1)	2001:db8:2:0::1/64
Cisco1841B	FastEthernet 0/0 (Fa0/0)	2001:db8:0:1::1/64
	Serial 0/0/0 (S0/0/0)	2001:db8:0:2::1/64
	Serial 0/0/1 (S0/0/1)	2001:db8:1:1::1/64
Cisco1841C	FastEthernet 0/0 (Fa0/0)	2001:db8:1:2::1/64
	Serial 0/0/0 (S0/0/0)	2001:db8:2:0::2/64
	Serial 0/0/1 (S0/0/1)	2001:db8:1:1::2/64

STEP1　ルーティングテーブルの確認

各ルーターのルーティングテーブルを確認します。

■ Cisco 1841A：

show ipv6 route コマンドで確認します。

```
Cisco1841A#show ipv6 route
IPv6 Routing Table - 11 entries
Codes: C - Connected, L - Local, S - Static, R - RIP, B - BGP
       U - Per-user Static route
```

```
        I1 - ISIS L1, I2 - ISIS L2, IA - ISIS interarea, IS - ISIS summary
        O - OSPF intra, OI - OSPF inter, OE1 - OSPF ext 1, OE2 - OSPF ext 2
        ON1 - OSPF NSSA ext 1, ON2 - OSPF NSSA ext 2
R   2001:DB8:0:1::/64 [120/2]
     via FE80::225:45FF:FEF7:BC80, Serial0/0/0
C   2001:DB8:0:2::/64 [0/0]
     via ::, Serial0/0/0
L   2001:DB8:0:2::2/128 [0/0]
     via ::, Serial0/0/0
C   2001:DB8:0:3::/64 [0/0]
     via ::, FastEthernet0/0
L   2001:DB8:0:3::1/128 [0/0]
     via ::, FastEthernet0/0
R   2001:DB8:1:1::/64 [120/2]
     via FE80::225:45FF:FEF7:BC80, Serial0/0/0
R   2001:DB8:1:2::/64 [120/3]
     via FE80::225:45FF:FEF7:BC80, Serial0/0/0
C   2001:DB8:2::/64 [0/0]
     via ::, Serial0/0/1
L   2001:DB8:2::1/128 [0/0]
     via ::, Serial0/0/1
L   FE80::/10 [0/0]
     via ::, Null0
L   FF00::/8 [0/0]
     via ::, Null0
Cisco1841A#
```

■ **Cisco1841B：**

show ipv6 route コマンドで確認します。

```
Cisco1841B#show ipv6 route
IPv6 Routing Table - 11 entries
Codes: C - Connected, L - Local, S - Static, R - RIP, B - BGP
       U - Per-user Static route
       I1 - ISIS L1, I2 - ISIS L2, IA - ISIS interarea, IS - ISIS summary
       O - OSPF intra, OI - OSPF inter, OE1 - OSPF ext 1, OE2 - OSPF ext 2
       ON1 - OSPF NSSA ext 1, ON2 - OSPF NSSA ext 2
C   2001:DB8:0:1::/64 [0/0]
     via ::, FastEthernet0/0
L   2001:DB8:0:1::1/128 [0/0]
     via ::, FastEthernet0/0
C   2001:DB8:0:2::/64 [0/0]
     via ::, Serial0/0/0
L   2001:DB8:0:2::1/128 [0/0]
     via ::, Serial0/0/0
R   2001:DB8:0:3::/64 [120/2]
     via FE80::223:EBFF:FE44:ED0C, Serial0/0/0
C   2001:DB8:1:1::/64 [0/0]
     via ::, Serial0/0/1
L   2001:DB8:1:1::1/128 [0/0]
     via ::, Serial0/0/1
R   2001:DB8:1:2::/64 [120/2]
     via FE80::222:55FF:FE52:B7AE, Serial0/0/1
R   2001:DB8:2::/64 [120/2]
     via FE80::223:EBFF:FE44:ED0C, Serial0/0/0
     via FE80::222:55FF:FE52:B7AE, Serial0/0/1
L   FE80::/10 [0/0]
     via ::, Null0
L   FF00::/8 [0/0]
```

```
           via ::, Null0
Cisco1841B#
```

■ Cisco1841C：

show ipv6 route コマンドで確認します。

```
Cisco1841C#show ipv6 route
IPv6 Routing Table - 11 entries
Codes: C - Connected, L - Local, S - Static, R - RIP, B - BGP
       U - Per-user Static route
       I1 - ISIS L1, I2 - ISIS L2, IA - ISIS interarea, IS - ISIS summary
       O - OSPF intra, OI - OSPF inter, OE1 - OSPF ext 1, OE2 - OSPF ext 2
       ON1 - OSPF NSSA ext 1, ON2 - OSPF NSSA ext 2
R   2001:DB8:0:1::/64 [120/2]
     via FE80::222:55FF:FE52:B23C, Serial0/0/1
R   2001:DB8:0:2::/64 [120/2]
     via FE80::223:EBFF:FE44:ED0C, Serial0/0/0
     via FE80::222:55FF:FE52:B23C, Serial0/0/1
R   2001:DB8:0:3::/64 [120/2]
     via FE80::223:EBFF:FE44:ED0C, Serial0/0/0
C   2001:DB8:1:1::/64 [0/0]
     via ::, Serial0/0/1
L   2001:DB8:1:1::2/128 [0/0]
     via ::, Serial0/0/1
C   2001:DB8:1:2::/64 [0/0]
     via ::, FastEthernet0/0
L   2001:DB8:1:2::1/128 [0/0]
     via ::, FastEthernet0/0
C   2001:DB8:2::/64 [0/0]
     via ::, Serial0/0/0
L   2001:DB8:2::2/128 [0/0]
     via ::, Serial0/0/0
L   FE80::/10 [0/0]
     via ::, Null0
L   FF00::/8 [0/0]
     via ::, Null0
Cisco1841C#
```

STEP2 デフォルトルートの伝搬

まずは Cisco1841A の FastEthernet 0/0 をデフォルトルートとみなし，その他のルーターにその情報を流す実習を行います。そのためにはまず静的に FastEthernet 0/0 の経路をデフォルトルートと設定しその後，その静的設定を RIPng に再配布する方法を取ります。

■ Cisco 1841A：

```
Cisco1841A(config)#ipv6 route ::/0 fastEthernet 0/0
Cisco1841A(config)#ipv6 router rip RIPng
Cisco1841A(config-rtr)#redistribute static
```

設定が終了したら，各ルーターのルーティングテーブルを確認します。

■ Cisco1841A：

show ipv6 route コマンドで確認します。

```
Cisco1841A#show ipv6 route
IPv6 Routing Table - 12 entries
```

Chapter 07 RIPngのデフォルトルートの伝搬

```
Codes: C - Connected, L - Local, S - Static, R - RIP, B - BGP
       U - Per-user Static route
       I1 - ISIS L1, I2 - ISIS L2, IA - ISIS interarea, IS - ISIS summary
       O - OSPF intra, OI - OSPF inter, OE1 - OSPF ext 1, OE2 - OSPF ext 2
       ON1 - OSPF NSSA ext 1, ON2 - OSPF NSSA ext 2
S   ::/0 [1/0]
     via ::, FastEthernet0/0
R   2001:DB8:0:1::/64 [120/2]
     via FE80::225:45FF:FEF7:BC80, Serial0/0/0
C   2001:DB8:0:2::/64 [0/0]
     via ::, Serial0/0/0
L   2001:DB8:0:2::2/128 [0/0]
     via ::, Serial0/0/0
C   2001:DB8:0:3::/64 [0/0]
     via ::, FastEthernet0/0
L   2001:DB8:0:3::1/128 [0/0]
     via ::, FastEthernet0/0
R   2001:DB8:1:1::/64 [120/2]
     via FE80::225:45FF:FEF7:BC80, Serial0/0/0
R   2001:DB8:1:2::/64 [120/3]
     via FE80::225:45FF:FEF7:BC80, Serial0/0/0
C   2001:DB8:2::/64 [0/0]
     via ::, Serial0/0/1
L   2001:DB8:2::1/128 [0/0]
     via ::, Serial0/0/1
L   FE80::/10 [0/0]
     via ::, Null0
L   FF00::/8 [0/0]
     via ::, Null0
Cisco1841A#
```

■ **Cisco1841B：**

show ipv6 route コマンドで確認します。

```
Cisco1841B#show ipv6 route
IPv6 Routing Table - 12 entries
Codes: C - Connected, L - Local, S - Static, R - RIP, B - BGP
       U - Per-user Static route
       I1 - ISIS L1, I2 - ISIS L2, IA - ISIS interarea, IS - ISIS summary
       O - OSPF intra, OI - OSPF inter, OE1 - OSPF ext 1, OE2 - OSPF ext 2
       ON1 - OSPF NSSA ext 1, ON2 - OSPF NSSA ext 2
R   ::/0 [120/2]
     via FE80::223:EBFF:FE44:ED0C, Serial0/0/0
C   2001:DB8:0:1::/64 [0/0]
     via ::, FastEthernet0/0
L   2001:DB8:0:1::1/128 [0/0]
     via ::, FastEthernet0/0
C   2001:DB8:0:2::/64 [0/0]
     via ::, Serial0/0/0
L   2001:DB8:0:2::1/128 [0/0]
     via ::, Serial0/0/0
R   2001:DB8:0:3::/64 [120/2]
     via FE80::223:EBFF:FE44:ED0C, Serial0/0/0
C   2001:DB8:1:1::/64 [0/0]
     via ::, Serial0/0/1
L   2001:DB8:1:1::1/128 [0/0]
     via ::, Serial0/0/1
R   2001:DB8:1:2::/64 [120/2]
     via FE80::222:55FF:FE52:B7AE, Serial0/0/1
```

```
R   2001:DB8:2::/64 [120/2]
     via FE80::223:EBFF:FE44:ED0C, Serial0/0/0
     via FE80::222:55FF:FE52:B7AE, Serial0/0/1
L   FE80::/10 [0/0]
     via ::, Null0
L   FF00::/8 [0/0]
     via ::, Null0
Cisco1841B#
```

■ Cisco1841C：

show ipv6 route コマンドで確認します。

```
Cisco1841C#show ipv6 route
IPv6 Routing Table - 12 entries
Codes: C - Connected, L - Local, S - Static, R - RIP, B - BGP
       U - Per-user Static route
       I1 - ISIS L1, I2 - ISIS L2, IA - ISIS interarea, IS - ISIS summary
       O - OSPF intra, OI - OSPF inter, OE1 - OSPF ext 1, OE2 - OSPF ext 2
       ON1 - OSPF NSSA ext 1, ON2 - OSPF NSSA ext 2
R   ::/0 [120/2]
     via FE80::223:EBFF:FE44:ED0C, Serial0/0/0
R   2001:DB8:0:1::/64 [120/2]
     via FE80::222:55FF:FE52:B23C, Serial0/0/1
R   2001:DB8:0:2::/64 [120/2]
     via FE80::222:55FF:FE52:B23C, Serial0/0/1
     via FE80::223:EBFF:FE44:ED0C, Serial0/0/0
R   2001:DB8:0:3::/64 [120/2]
     via FE80::223:EBFF:FE44:ED0C, Serial0/0/0
C   2001:DB8:1:1::/64 [0/0]
     via ::, Serial0/0/1
L   2001:DB8:1:1::2/128 [0/0]
     via ::, Serial0/0/1
C   2001:DB8:1:2::/64 [0/0]
     via ::, FastEthernet0/0
L   2001:DB8:1:2::1/128 [0/0]
     via ::, FastEthernet0/0
C   2001:DB8:2::/64 [0/0]
     via ::, Serial0/0/0
L   2001:DB8:2::2/128 [0/0]
     via ::, Serial0/0/0
L   FE80::/10 [0/0]
     via ::, Null0
L   FF00::/8 [0/0]
     via ::, Null0
Cisco1841C#
```

Chapter 07 RIPngのデフォルトルートの伝搬

■実習終了時設定内容

■Cisco1814A：

```
!
version 12.4
service timestamps debug datetime msec
service timestamps log datetime msec
no service password-encryption
!
hostname Cisco1841A
!
boot-start-marker
boot-end-marker
!
logging message-counter syslog
!
no aaa new-model
dot11 syslog
ip source-route
!
!
!
!
ip cef
ipv6 unicast-routing
ipv6 cef
!
multilink bundle-name authenticated
!
!
!
!
!
archive
 log config
  hidekeys
!
!
!
!
!
!
!
interface FastEthernet0/0
 no ip address
 duplex auto
 speed auto
 ipv6 address 2001:DB8:0:3::1/64
 ipv6 enable
 no keepalive
!
interface FastEthernet0/1
 no ip address
 shutdown
 duplex auto
 speed auto
!
interface Serial0/0/0
```

```
 no ip address
 ipv6 address 2001:DB8:0:2::2/64
 ipv6 enable
 ipv6 rip RIPng enable
 clock rate 125000
!
interface Serial0/0/1
 no ip address
 ipv6 address 2001:DB8:2::1/64
 ipv6 enable
 clock rate 125000
!
ip forward-protocol nd
!
!
ip http server
no ip http secure-server
!
ipv6 route ::/0 FastEthernet0/0
ipv6 router rip RIPng
 redistribute connected
 redistribute static
!
!
!
!
control-plane
!
!
!
line con 0
line aux 0
line vty 0 4
 login
!
scheduler allocate 20000 1000
end
```

■ Cisco1841B：

```
!
version 12.4
service timestamps debug datetime msec
service timestamps log datetime msec
no service password-encryption
!
hostname Cisco1841B
!
boot-start-marker
boot-end-marker
!
!
no aaa new-model
ip cef
!
!
!
!
ip auth-proxy max-nodata-conns 3
ip admission max-nodata-conns 3
```

Chapter 07 RIPngのデフォルトルートの伝搬

```
!
ipv6 unicast-routing
!
!
!
!
!
!
!
interface FastEthernet0/0
 no ip address
 duplex auto
 speed auto
 ipv6 address 2001:DB8:0:1::1/64
 ipv6 enable
 no keepalive
!
interface FastEthernet0/1
 no ip address
 shutdown
 duplex auto
 speed auto
!
!
interface Serial0/0/0
 no ip address
 ipv6 address 2001:DB8:0:2::1/64
 ipv6 enable
 ipv6 rip RIPng enable
 no fair-queue
!
interface Serial0/0/1
 no ip address
 ipv6 address 2001:DB8:1:1::1/64
 ipv6 rip RIPng enable
!
ip forward-protocol nd
!
!
ip http server
no ip http secure-server
!
ipv6 router rip RIPng
 redistribute connected
!
!
!
!
control-plane
!
!
!
line con 0
line aux 0
line vty 0 4
 login
!
scheduler allocate 20000 1000
end
```

■ Cisco1841C：

```
!
version 12.4
service timestamps debug datetime msec
service timestamps log datetime msec
no service password-encryption
!
hostname Cisco1841C
!
boot-start-marker
boot-end-marker
!
!
no aaa new-model
ip cef
!
!
!
!
ip auth-proxy max-nodata-conns 3
ip admission max-nodata-conns 3
!
ipv6 unicast-routing
!
!
!
!
!
!
!
interface FastEthernet0/0
 no ip address
 duplex auto
 speed auto
 ipv6 address 2001:DB8:1:2::1/64
 ipv6 enable
 no keepalive
!
interface FastEthernet0/1
 no ip address
 shutdown
 duplex auto
 speed auto
!
interface Serial0/0/0
 no ip address
 ipv6 address 2001:DB8:2::2/64
 ipv6 enable
 ipv6 rip RIPng enable
!
interface Serial0/0/1
 no ip address
 ipv6 address 2001:DB8:1:1::2/64
 ipv6 enable
 ipv6 rip RIPng enable
 clock rate 125000
!
ip forward-protocol nd
```

```
!
!
ip http server
no ip http secure-server
!
ipv6 router rip RIPng
 redistribute connected
!
!
!
!
control-plane
!
!
!
line con 0
line aux 0
line vty 0 4
 login
!
scheduler allocate 20000 1000
end
```

Chapter 08

OSPFv3の設定

08-1 マルチベンダー機器による
　　　OSPFv3基本設定

08-2 シスコ機器による
　　　OSPFv3基本設定

Chapter 08 OSPFv3の設定

08-1 マルチベンダー機器による OSPFv3基本設定

　IPv6を用いたネットワーク上でOSPFv3を動作させる際に必要な基本的な項目について実習を行います。なお本実習では1台のCiscoルーター（Cisco1812J）と2台のAX620Rを用います。

■実習トポロジー

この節で利用するトポロジーは以下のとおりです。

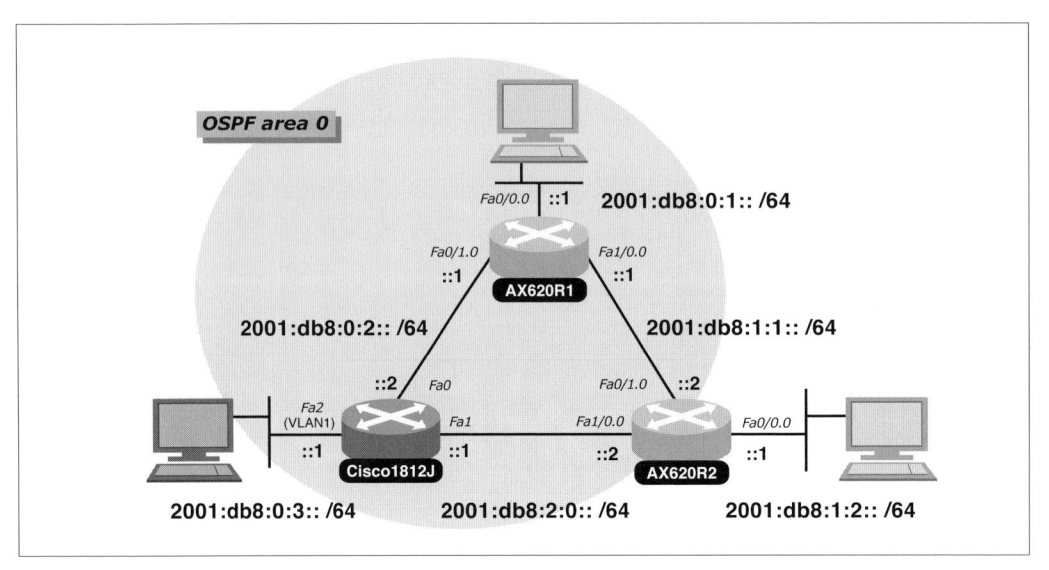

図8-1　ネットワークトポロジー

■インターフェース情報

表8-1　インターフェースアドレス割当一覧

ルーター	インターフェース	IPv6アドレス
Cisco1812J	FastEthernet 0 (Fa0)	2001:db8:0:2::2/64
	FastEthernet 1 (Fa1)	2001:db8:2:0::1/64
	FastEthernet 2 (VLAN1) (Fa2)	2001:db8:0:3::1/64
AX620R1	FastEthernet 0/0.0 (Fa0/0.0)	2001:db8:0:1::1/64
	FastEthernet 0/1.0 (Fa0/1.0)	2001:db8:0:2::1/64
	FastEthernet 1/0.0 (Fa1/0.0)	2001:db8:1:1::1/64
AX620R2	FastEthernet 0/0.0 (Fa0/0.0)	2001:db8:1:2::1/64
	FastEthernet 0/1.0 (Fa0/1.0)	2001:db8:1:1::2/64
	FastEthernet 1/0.0 (Fa1/0.0)	2001:db8:2:0::2/64

■ ルーター ID 情報

表8-2　ルーターID情報一覧

ルーター	ルーターID
Cisco1812J	10.0.0.1
AX620R1	172.16.1.1
AX620R2	192.168.1.1

■ 設定手順

STEP1　物理接続の実施

各ルーターおよび PC を "図 8-1　ネットワークトポロジー " のとおりに接続します。

STEP2　ホスト名の設定と IPv6 の有効化

ルーターのホスト名と Cisco ルーターに IPv6 トラフィック転送有効化の設定を行います。

■ Cisco1812J：

```
Router(config)#hostname Cisco1812J
Cisco1812J(config)#ipv6 unicast-routing
```

■ AX620R1：

```
Router(config)#hostname AX620R1
```

■ AX620R2：

```
Router(config)#hostname AX620R2
```

STEP3　インターフェースへの IPv6 アドレスの設定

すべてのルーターに IPv6 アドレスを設定し，インターフェースを有効化します。アドレスは "表 8-1　インターフェースアドレス割当一覧 " を参照してください。なお，Cisco1812J は FastEthernet 2 から 6 までスイッチポートとして動作するため，VLAN1 として設定します。

■ Cisco1812J：

```
Cisco1812J(config)#interface fastethernet0
Cisco1812J(config-if)#ipv6 enable
Cisco1812J(config-if)#ipv6 address 2001:db8:0:2::2/64
Cisco1812J(config-if)#no shutdown
Cisco1812J(config)#interface fastethernet1
Cisco1812J(config-if)#ipv6 enable
Cisco1812J(config-if)#ipv6 address 2001:db8:2:0::1/64
Cisco1812J(config-if)#no shutdown
Cisco1812J(config)#interface vlan1
Cisco1812J(config-if)#ipv6 enable
Cisco1812J(config-if)#ipv6 address 2001:db8:0:3::1/64
Cisco1812J(config-if)#no shutdown
```

Chapter 08 OSPFv3の設定

■ AX620R1：

```
AX620R1(config)#interface FastEthernet0/0.0
AX620R1(config-FastEthernet0/0.0)#ipv6 enable
AX620R1(config-FastEthernet0/0.0)#ipv6 address 2001:db8:0:1::1/64
AX620R1(config-FastEthernet0/0.0)#no shutdown
AX620R1(config)#interface FastEthernet0/1.0
AX620R1(config-FastEthernet0/1.0)#ipv6 enable
AX620R1(config-FastEthernet0/1.0)#ipv6 address 2001:db8:0:2::1/64
AX620R1(config-FastEthernet0/1.0)#no shutdown
AX620R1(config)#interface FastEthernet1/0.0
AX620R1(config-FastEthernet1/0.0)#ipv6 enable
AX620R1(config-FastEthernet1/0.0)#ipv6 address 2001:db8:1:1::1/64
AX620R1(config-FastEthernet1/0.0)#no shutdown
```

■ AX620R2：

```
AX620R2(config)#interface FastEthernet0/0.0
AX620R2(config-FastEthernet0/0.0)#ipv6 enable
AX620R2(config-FastEthernet0/0.0)#ipv6 address 2001:db8:1:2::1/64
AX620R2(config-FastEthernet0/0.0)#no shutdown
AX620R2(config)#interface FastEthernet1/0.0
AX620R2(config-FastEthernet1/0.0)#ipv6 enable
AX620R2(config-FastEthernet1/0.0)#ipv6 address 2001:db8:2:0::2/64
AX620R2(config-FastEthernet1/0.0)#no shutdown
AX620R2(config)#interface FastEthernet0/1.0
AX620R2(config-FastEthernet0/1.0)#ipv6 enable
AX620R2(config-FastEthernet0/1.0)#ipv6 address 2001:db8:1:1::2/64
AX620R2(config-FastEthernet0/1.0)#no shutdown
```

STEP4 インターフェースのアドレス確認

インターフェースのステータスと動作していることを確認します。

show ipv6 interface brief または **show ipv6 address** コマンドを用いてインターフェース情報を表示させます。下記に出力を示します。インターフェース名の後の表示が [up/up] なら動作しています。

■ Cisco1812J：

show ipv6 interface brief コマンドで確認します。

```
Cisco1812J#show ipv6 interface brief
FastEthernet0              [up/up]
    FE80::226:BFF:FEF6:B5FA
    2001:DB8:0:2::2
FastEthernet1              [up/up]
    FE80::226:BFF:FEF6:B5FB
    2001:DB8:2::1
BRI0                       [administratively down/down]
BRI0:1                     [administratively down/down]
BRI0:2                     [administratively down/down]
FastEthernet2              [up/up]
FastEthernet3              [up/down]
FastEthernet4              [up/down]
FastEthernet5              [up/down]
FastEthernet6              [up/down]
FastEthernet7              [up/down]
FastEthernet8              [up/down]
FastEthernet9              [down/down]
```

```
Vlan1                              [up/up]
   FE80::226:BFF:FEF6:B5FA
   2001:DB8:0:3::1
```

■ **AX620R1：**

show ipv6 address コマンドで確認します。

```
AX620R1(config)#show ipv6 address
Interface FastEthernet0/0.0 is up, line protocol is up
  Global address(es):
    2001:db8:0:1::1 prefixlen 64
    2001:db8:0:1::0 prefixlen 64 anycast
  Link-local address(es):
    fe80::260:b9ff:fe4b:a477 prefixlen 64
    fe80::0 prefixlen 64 anycast
  Multicast address(es):
    ff02::1
    ff02::2
    ff02::5
    ff02::6
    ff02::1:ff00:0
    ff02::1:ff00:1
    ff02::1:ff4b:a477
Interface FastEthernet0/1.0 is up, line protocol is up
  Global address(es):
    2001:db8:0:2::1 prefixlen 64
    2001:db8:0:2::0 prefixlen 64 anycast
  Link-local address(es):
    fe80::260:b9ff:fe4b:a4f7 prefixlen 64
    fe80::0 prefixlen 64 anycast
  Multicast address(es):
    ff02::1
    ff02::2
    ff02::5
    ff02::6
    ff02::1:ff00:0
    ff02::1:ff00:1
    ff02::1:ff4b:a4f7
Interface FastEthernet1/0.0 is up, line protocol is up
  Global address(es):
    2001:db8:1:1::1 prefixlen 64
    2001:db8:1:1::0 prefixlen 64 anycast
  Link-local address(es):
    fe80::260:b9ff:fe4b:a40f prefixlen 64
    fe80::0 prefixlen 64 anycast
  Multicast address(es):
    ff02::1
    ff02::2
    ff02::5
    ff02::6
    ff02::1:ff00:0
    ff02::1:ff00:1
    ff02::1:ff4b:a40f
Interface Loopback0.0 is up, line protocol is up
  Orphan address(es):
    ::1 prefixlen 128
Interface Loopback1.0 is up, line protocol is up
Interface Null0.0 is up, line protocol is up
Interface Null1.0 is up, line protocol is up
```

Chapter 08 OSPFv3の設定

■ **AX620R2：**

show ipv6 address コマンドで確認します。

```
AX620R2(config)#show ipv6 address
Interface FastEthernet0/0.0 is up, line protocol is up
  Global address(es):
    2001:db8:1:2::1 prefixlen 64
    2001:db8:1:2:: prefixlen 64 anycast
  Link-local address(es):
    fe80::260:b9ff:fe4f:225f prefixlen 64
    fe80:: prefixlen 64 anycast
  Multicast address(es):
    ff02::1
    ff02::2
    ff02::5
    ff02::6
    ff02::1:ff00:0
    ff02::1:ff00:1
    ff02::1:ff4f:225f
Interface FastEthernet0/1.0 is up, line protocol is up
  Global address(es):
    2001:db8:1:1::2 prefixlen 64
    2001:db8:1:1:: prefixlen 64 anycast
  Link-local address(es):
    fe80::260:b9ff:fe4f:22df prefixlen 64
    fe80:: prefixlen 64 anycast
  Multicast address(es):
    ff02::1
    ff02::2
    ff02::5
    ff02::6
    ff02::1:ff00:0
    ff02::1:ff00:2
    ff02::1:ff4f:22df
Interface FastEthernet1/0.0 is up, line protocol is up
  Global address(es):
    2001:db8:2::2 prefixlen 64
    2001:db8:2:: prefixlen 64 anycast
  Link-local address(es):
    fe80::260:b9ff:fe4f:223f prefixlen 64
    fe80:: prefixlen 64 anycast
  Multicast address(es):
    ff02::1
    ff02::2
    ff02::5
    ff02::6
    ff02::1:ff00:0
    ff02::1:ff00:2
    ff02::1:ff4f:223f
Interface Loopback0.0 is up, line protocol is up
  Orphan address(es):
    ::1 prefixlen 128
Interface Loopback1.0 is up, line protocol is up
Interface Null0.0 is up, line protocol is up
Interface Null1.0 is up, line protocol is up
```

STEP5 OSPFv3 プロセスの設定

OSPFv3 をプロセス番号 1 で起動し，各ルーターにルーター ID を設定します。ルーター ID は " 表 8-2　ルーター ID 情報一覧 " を参照してください。

■ Cisco1812J：

```
Cisco1812J(config)#ipv6 router ospf 1
Cisco1812J(config-router)#router-id 10.0.0.1
```

■ AX620R1：

```
AX620R1(config)#ipv6 router ospf 1
AX620R1(config -ospfv3-1)#router-id 172.16.1.1
AX620R2(config- ospfv3-1)#clear ipv6 ospf process
```

■ AX620R2：

```
AX620R2(config)#ipv6 router ospf 1
AX620R2(config- ospfv3-1)#router-id 192.168.1.1
AX620R2(config- ospfv3-1)#clear ipv6 ospf process
```

STEP6 インターフェースへの OSPFv3 プロセス及びエリアの割当

各インターフェースに OSPFv3 プロセス 1，エリア 0 を割り当てます。

■ Cisco1812J：

```
Cisco1812J(config)#interface FastEthernet0
Cisco1812J(config-if)#ipv6 ospf 1 area 0
Cisco1812J(config)#interface FastEthernet1
Cisco1812J(config-if)#ipv6 ospf 1 area 0
Cisco1812J(config)#interface vlan1
Cisco1812J(config-if)#ipv6 ospf 1 area 0
```

■ AX620R1：

```
AX620R1(config-ospfv3-1)#network FastEthernet0/0.0 area 0
AX620R1(config-ospfv3-1)#network FastEthernet0/1.0 area 0
AX620R1(config-ospfv3-1)#network FastEthernet1/0.0 area 0
```

■ AX620R2：

```
AX620R2(config-ospfv3-1)#network FastEthernet0/0.0 area 0
AX620R2(config-ospfv3-1)#network FastEthernet0/1.0 area 0
AX620R2(config-ospfv3-1)#network FastEthernet1/0.0 area 0
```

STEP7 OSPFv3 設定確認

各ルーターで動作しているルーティングプロトコルが設定した内容 (OSPFv3 process1 Area0) であることを **show ipv6 protocols** コマンドで確認します。

■ Cisco1812J：

show ipv6 protocols コマンドで確認します。

```
Cisco1812J#show ipv6 protocols
IPv6 Routing Protocol is "connected"
IPv6 Routing Protocol is "static"
```

Chapter 08 OSPFv3の設定

```
IPv6 Routing Protocol is "ospf 1"
  Interfaces (Area 0):
    Vlan1
    FastEthernet1
    FastEthernet0
  Redistribution:
    None
```

■ **AX620R1：**

show ipv6 protocols コマンドで確認します。

```
AX620R1(config)#show ipv6 protocols
IPv6 unicast routing is enabled
  Process switching queue-len: 1/3 (last/peak), overflows: 0
  Host transit queue-len: 3/5 (last/peak), overflows: 0
  Fragment transit queue-len: 0/0 (last/peak), overflows: 0
  FIB entries:
    System: 40/unlimited (busy/max)
    Dynamic routing: 7/unlimited (busy/max)
  Routing cache bucket(s): 0/4096/0/4096 (busy/free/garbage/max)
  Maximum path(s): 16
  Load sharing algorithm is per-packet round robin
IPv6 multicast routing is disabled
IPv6 source-routing type-0 is disabled
  Received 0 type-0, 0 other types
  Discarded 0 type-0, 0 other types
    0 truncated, 0 invalid address, 0 other error
IPv6 unified forwarding service cache is disabled
IPv6 reassembly service is enabled
  Reassemble buffer size is 0/65535 octets (peak/max)
  Reassemble buffers: 0/0/16 (curr/peak/max)
ICMP for IPv6 is enabled
  Error message limited to one every 1000 milliseconds
Neighbor discovery for IPv6 is enabled
  Neighbor cache(s): 1/1024 (busy/max)
Path MTU discovery is enabled
  Cache entries: 0/0/unlimited (curr/peak/max)
  System errors: 0/0 (overflows/alloc fails)
  Too big received: 0/0/0 (received/under 1280/over 65535)
IPv6 routing protocol is "ospf 1"
  Router ID 172.16.1.1
  RIB entries: 7
  Autonomous system boundary capability: no
  Area border capability: no
  Number of areas in this router: 1
  Originating default route is disabled
  Administrative distance:
    External 110, Intra-Area 110, Inter-Area 110
  Delay time between receiving a change to SPF calculations: 5 seconds
  Hold time between consecutive SPF calculations: 10 seconds
  Redistribution:
    None
  Passive interface(s):
    None
  Network(s):
    FastEthernet0/0.0, area 0(0.0.0.0)
    FastEthernet0/1.0, area 0(0.0.0.0)
    FastEthernet1/0.0, area 0(0.0.0.0)
```

■ AX620R2:

`show ipv6 protocols` コマンドで確認します。

```
AX620R2(config)#show ipv6 protocols
IPv6 unicast routing is enabled
  Process switching queue-len: 1/3 (last/peak), overflows: 0
  Host transit queue-len: 3/4 (last/peak), overflows: 0
  Fragment transit queue-len: 0/0 (last/peak), overflows: 0
  FIB entries:
    System: 40/unlimited (busy/max)
    Dynamic routing: 7/unlimited (busy/max)
  Routing cache bucket(s): 0/4096/0/4096 (busy/free/garbage/max)
  Maximum path(s): 16
  Load sharing algorithm is per-packet round robin
IPv6 multicast routing is disabled
IPv6 source-routing type-0 is disabled
  Received 0 type-0, 0 other types
  Discarded 0 type-0, 0 other types
    0 truncated, 0 invalid address, 0 other error
IPv6 unified forwarding service cache is disabled
IPv6 reassembly service is enabled
  Reassemble buffer size is 0/65535 octets (peak/max)
  Reassemble buffers: 0/0/16 (curr/peak/max)
ICMP for IPv6 is enabled
  Error message limited to one every 1000 milliseconds
Neighbor discovery for IPv6 is enabled
  Neighbor cache(s): 1/1024 (busy/max)
Path MTU discovery is enabled
  Cache entries: 0/0/unlimited (curr/peak/max)
  System errors: 0/0 (overflows/alloc fails)
  Too big received: 0/0/0 (received/under 1280/over 65535)
IPv6 routing protocol is "ospf 1"
  Router ID 192.168.1.1
  RIB entries: 7
  Autonomous system boundary capability: no
  Area border capability: no
  Number of areas in this router: 1
  Originating default route is disabled
  Administrative distance:
    External 110, Intra-Area 110, Inter-Area 110
  Delay time between receiving a change to SPF calculations: 5 seconds
  Hold time between consecutive SPF calculations: 10 seconds
  Redistribution:
    None
  Passive interface(s):
    None
  Network(s):
    FastEthernet0/0.0, area 0(0.0.0.0)
    FastEthernet0/1.0, area 0(0.0.0.0)
    FastEthernet1/0.0, area 0(0.0.0.0)
```

STEP8 ルーター相互の隣接情報確認

各ルーターの隣接情報などを `show ipv6 ospf neighbor` コマンドで確認します。

■ Cisco1812J:

`show ipv6 ospf neighbor` コマンドで確認します。

Chapter 08 OSPFv3の設定

```
Cisco1812J#show ipv6 ospf neighbor

Neighbor ID     Pri   State         Dead Time    Interface ID    Interface
192.168.1.1      1    FULL/BDR      00:00:34     3               FastEthernet1
172.16.1.1       1    FULL/BDR      00:00:31     2               FastEthernet0
```

■ AX620R1：

show ipv6 ospf neighbor コマンドで確認します。

```
AX620R1(config)#show ipv6 ospf neighbor
Neighbor ID     PID   Pri   State         Age    Uptime      Interface
10.0.0.1         1     1    FULL/DR       2      0:17:00     FastEthernet0/1.0
192.168.1.1      1     1    FULL/DR       6      0:16:57     FastEthernet1/0.0
```

■ AX620R2：

show ipv6 ospf neighbor コマンドで確認します。

```
AX620R2(config)#show ipv6 ospf neighbor
Neighbor ID     PID   Pri   State         Age    Uptime      Interface
172.16.1.1       1     1    FULL/BDR      1      0:22:28     FastEthernet0/1.0
10.0.0.1         1     1    FULL/DR       5      0:51:17     FastEthernet1/0.0
```

STEP9　OSPFv3 データベースの確認

さらに各ルーターの OSPFv3 データベースの内容を **show ipv6 ospf database** コマンドで確認します。

■ Cisco1812J：

show ipv6 ospf database コマンドで確認します。

```
Cisco1812J#show ipv6 ospf database

            OSPFv3 Router with ID (10.0.0.1) (Process ID 1)

                Router Link States (Area 0)

ADV Router          Age         Seq#            Fragment ID    Link count    Bits
10.0.0.1            293         0x8000002B      0              2             None
172.16.1.1          740         0x80000049      0              2             None
192.168.1.1         776         0x80000026      0              2             None

                Net Link States (Area 0)

ADV Router          Age         Seq#            Link ID    Rtr count
10.0.0.1            784         0x80000001      2          2
10.0.0.1            1569        0x80000002      3          2
192.168.1.1         776         0x8000000B      2          2

            Link (Type-8) Link States (Area 0)

ADV Router          Age         Seq#            Link ID    Interface
10.0.0.1            328         0x80000002      20         Vl1
172.16.1.1          3568        0x80000003      1          Vl1
192.168.1.1         3487        0x80000002      1          Vl1
10.0.0.1            335         0x80000007      3          Fa1
192.168.1.1         713         0x80000004      3          Fa1
```

```
            10.0.0.1             336              0x80000006  2              Fa0
            172.16.1.1           786              0x80000005  2              Fa0


                        Intra Area Prefix Link States (Area 0)

ADV Router           Age            Seq#              Link ID     Ref-lstype    Ref-LSID
10.0.0.1             329            0x80000001        0           0x2001        0
10.0.0.1             785            0x80000001        2048        0x2002        2
10.0.0.1             1570           0x80000002        3072        0x2002        3
172.16.1.1           742            0x80000031        0           0x2001        0
192.168.1.1          778            0x80000038        0           0x2001        0
192.168.1.1          710            0x80000013        2           0x2002        2
```

■ **AX620R1：**

show ipv6 ospf database コマンドで確認します。

```
AX620R1(config)#show ipv6 ospf database

OSPFv3 router with process ID 1
Router ID 172.16.1.1

  Link LSAs for Interface FastEthernet0/0.0(ID 1)
  Adv. router       Age       Seq. No.       LSID       Priority    Prefixes
  172.16.1.1        998       0x80000003     1          1           1

  Link LSAs for Interface FastEthernet0/1.0(ID 2)
  Adv. router       Age       Seq. No.       LSID       Priority    Prefixes
  10.0.0.1          590       0x80000006     2          1           1
  172.16.1.1        1038      0x80000005     2          1           1

  Link LSAs for Interface FastEthernet1/0.0(ID 3)
  Adv. router       Age       Seq. No.       LSID       Priority    Prefixes
  172.16.1.1        1035      0x80000003     3          1           1
  192.168.1.1       963       0x80000004     2          1           1

  Router LSAs for Area 0.0.0.0(0)
  Adv. router       Age       Seq. No.       LSID       Link        Flags
  10.0.0.1          548       0x8000002b     0          2
  172.16.1.1        998       0x80000049     0          2
  192.168.1.1       1031      0x80000026     0          2

  Network LSAs for Area 0.0.0.0(0)
  Adv. router       Age       Seq. No.       LSID       Routers
  10.0.0.1          1040      0x80000001     2          2
  10.0.0.1          1824      0x80000002     3          2
  192.168.1.1       1031      0x8000000b     2          2

  Intra-Area-Prefix LSAs for Area 0.0.0.0(0)
  Adv. router       Age       Seq. No.       LSID       Prefixes    Reftype    RefID
  10.0.0.1          585       0x80000001     0          1           Router     0
  10.0.0.1          1041      0x80000001     2048       1           Network    2
  10.0.0.1          1825      0x80000002     3072       1           Network    3
  172.16.1.1        1000      0x80000031     0          1           Router     0
  192.168.1.1       1032      0x80000038     0          1           Router     0
  192.168.1.1       965       0x80000013     2          1           Network    2
```

■ AX620R2：

show ipv6 ospf database コマンドで確認します。

```
AX620R2(config)#show ipv6 ospf database

OSPFv3 router with process ID 1
Router ID 192.168.1.1

   Link LSAs for Interface FastEthernet0/0.0(ID 1)
   Adv. router        Age    Seq. No.       LSID       Priority    Prefixes
   192.168.1.1        1292   0x80000004     1          1           1

   Link LSAs for Interface FastEthernet0/1.0(ID 2)
   Adv. router        Age    Seq. No.       LSID       Priority    Prefixes
   172.16.1.1         1404   0x80000003     3          1           1
   192.168.1.1        1329   0x80000004     2          1           1

   Link LSAs for Interface FastEthernet1/0.0(ID 3)
   Adv. router        Age    Seq. No.       LSID       Priority    Prefixes
   10.0.0.1           958    0x80000007     3          1           1
   192.168.1.1        1333   0x80000004     3          1           1

   Router LSAs for Area 0.0.0.0(0)
   Adv. router        Age    Seq. No.       LSID       Link        Flags
   10.0.0.1           916    0x8000002b     0          2
   172.16.1.1         1367   0x80000049     0          2
   192.168.1.1        1399   0x80000026     0          2

   Network LSAs for Area 0.0.0.0(0)
   Adv. router        Age    Seq. No.       LSID       Routers
   10.0.0.1           1408   0x80000001     2          2
   10.0.0.1           213    0x80000003     3          2
   192.168.1.1        1399   0x8000000b     2          2

   Intra-Area-Prefix LSAs for Area 0.0.0.0(0)
   Adv. router        Age    Seq. No.       LSID       Prefixes    Reftype     RefID
   10.0.0.1           952    0x80000001     0          1           Router      0
   10.0.0.1           1408   0x80000001     2048       1           Network     2
   10.0.0.1           213    0x80000003     3072       1           Network     3
   172.16.1.1         1368   0x80000031     0          1           Router      0
   192.168.1.1        1399   0x80000038     0          1           Router      0
   192.168.1.1        1331   0x80000013     2          1           Network     2
```

STEP10　ルーティングテーブルの確認

show ipv6 route コマンドを用いて，ルーティングテーブル情報を表示します。下記に出力例を示します。

■ Cisco1812J：

show ipv6 route コマンドで確認します。

```
Cisco1812J#show ipv6 route
IPv6 Routing Table - 10 entries
Codes: C - Connected, L - Local, S - Static, R - RIP, B - BGP
       U - Per-user Static route, M - MIPv6
       I1 - ISIS L1, I2 - ISIS L2, IA - ISIS interarea, IS - ISIS summary
       O - OSPF intra, OI - OSPF inter, OE1 - OSPF ext 1, OE2 - OSPF ext 2
```

```
            ON1 - OSPF NSSA ext 1, ON2 - OSPF NSSA ext 2
            D - EIGRP, EX - EIGRP external
O    2001:DB8:0:1::/64 [110/2]
        via FE80::260:B9FF:FE4B:A4F7, FastEthernet0
C    2001:DB8:0:2::/64 [0/0]
        via ::, FastEthernet0
L    2001:DB8:0:2::2/128 [0/0]
        via ::, FastEthernet0
C    2001:DB8:0:3::/64 [0/0]
        via ::, Vlan1
L    2001:DB8:0:3::1/128 [0/0]
        via ::, Vlan1
O    2001:DB8:1:1::/64 [110/2]
        via FE80::260:B9FF:FE4F:223F, FastEthernet1
        via FE80::260:B9FF:FE4B:A4F7, FastEthernet0
O    2001:DB8:1:2::/64 [110/2]
        via FE80::260:B9FF:FE4F:223F, FastEthernet1
C    2001:DB8:2::/64 [0/0]
        via ::, FastEthernet1
L    2001:DB8:2::1/128 [0/0]
        via ::, FastEthernet1
L    FF00::/8 [0/0]
        via ::, Null0
```

■ AX620R1：

`show ipv6 route` コマンドで確認します。

```
AX620R1(config)#show ipv6 route
IPv6 Routing Table - 16 entries, unlimited
Codes: C - Connected, L - Local, S - Static
       R - RIPng, O - OSPF, IA - OSPF inter area
       E1 - OSPF external type 1, E2 - OSPF external type 2, B - BGP
       s - Summary
Timers: Uptime/Age
C    2001:db8:0:1::/64 global [0/1]
        via ::, FastEthernet0/0.0, 0:41:46/0:00:00
O    2001:db8:0:1::/64 global [110/1]
        via ::, FastEthernet0/0.0, 0:09:52/0:00:00
L    2001:db8:0:1::0/128 global [0/1]
        via ::, FastEthernet0/0.0, 0:41:48/0:00:00
L    2001:db8:0:1::1/128 global [0/1]
        via ::, FastEthernet0/0.0, 0:41:47/0:00:00
C    2001:db8:0:2::/64 global [0/1]
        via ::, FastEthernet0/1.0, 0:41:47/0:00:00
O    2001:db8:0:2::/64 global [110/1]
        via ::, FastEthernet0/1.0, 0:09:53/0:00:00
L    2001:db8:0:2::0/128 global [0/1]
        via ::, FastEthernet0/1.0, 0:41:48/0:00:00
L    2001:db8:0:2::1/128 global [0/1]
        via ::, FastEthernet0/1.0, 0:41:47/0:00:00
O    2001:db8:0:3::/64 global [110/2]
        via fe80::226:bff:fef6:b5fa, FastEthernet0/1.0, 0:09:54/0:00:00
C    2001:db8:1:1::/64 global [0/1]
        via ::, FastEthernet1/0.0, 0:41:48/0:00:00
O    2001:db8:1:1::/64 global [110/1]
        via ::, FastEthernet1/0.0, 0:09:54/0:00:00
L    2001:db8:1:1::0/128 global [0/1]
        via ::, FastEthernet1/0.0, 0:41:49/0:00:00
L    2001:db8:1:1::1/128 global [0/1]
```

Chapter 08 OSPFv3の設定

```
                via ::, FastEthernet1/0.0, 0:41:48/0:00:00
O       2001:db8:1:2::/64 global [110/2]
                via fe80::260:b9ff:fe4f:22df, FastEthernet1/0.0, 0:09:54/0:00:00
O       2001:db8:2::/64 global [110/2]
                via fe80::226:bff:fef6:b5fa, FastEthernet0/1.0, 0:09:54/0:00:00
                via fe80::260:b9ff:fe4f:22df, FastEthernet1/0.0, 0:09:54/0:00:00
```

■ AX620R2：

show ipv6 route コマンドで確認します。

```
AX620R2(config)#show ipv6 route
IPv6 Routing Table - 16 entries, unlimited
Codes: C - Connected, L - Local, S - Static
       R - RIPng, O - OSPF, IA - OSPF inter area
       E1 - OSPF external type 1, E2 - OSPF external type 2, B - BGP
       s - Summary
Timers: Uptime/Age
O       2001:db8:0:1::/64 global [110/2]
                via fe80::260:b9ff:fe4b:a40f, FastEthernet0/1.0, 0:04:05/0:00:00
O       2001:db8:0:2::/64 global [110/2]
                via fe80::226:bff:fef6:b5fb, FastEthernet1/0.0, 0:04:05/0:00:00
                via fe80::260:b9ff:fe4b:a40f, FastEthernet0/1.0, 0:04:05/0:00:00
O       2001:db8:0:3::/64 global [110/2]
                via fe80::226:bff:fef6:b5fb, FastEthernet1/0.0, 0:04:05/0:00:00
C       2001:db8:1:1::/64 global [0/1]
                via ::, FastEthernet0/1.0, 1:22:24/0:00:00
O       2001:db8:1:1::/64 global [110/1]
                via ::, FastEthernet0/1.0, 0:04:06/0:00:00
L       2001:db8:1:1::/128 global [0/1]
                via ::, FastEthernet0/1.0, 1:22:26/0:00:00
L       2001:db8:1:1::2/128 global [0/1]
                via ::, FastEthernet0/1.0, 1:22:25/0:00:00
C       2001:db8:1:2::/64 global [0/1]
                via ::, FastEthernet0/0.0, 1:23:04/0:00:00
O       2001:db8:1:2::/64 global [110/1]
                via ::, FastEthernet0/0.0, 0:04:06/0:00:00
L       2001:db8:1:2::/128 global [0/1]
                via ::, FastEthernet0/0.0, 1:23:05/0:00:00
L       2001:db8:1:2::1/128 global [0/1]
                via ::, FastEthernet0/0.0, 1:23:05/0:00:00
C       2001:db8:2::/64 global [0/1]
                via ::, FastEthernet1/0.0, 1:22:44/0:00:00
O       2001:db8:2::/64 global [110/1]
                via ::, FastEthernet1/0.0, 0:04:07/0:00:00
L       2001:db8:2::/128 global [0/1]
                via ::, FastEthernet1/0.0, 1:22:45/0:00:00
L       2001:db8:2::2/128 global [0/1]
                via ::, FastEthernet1/0.0, 1:22:44/0:00:00
```

STEP11　通信による確認

各ルーターから直接つながっていない別のルーターのインターフェースまでの通信を **ping** コマンドで確認します。

■ Cisco1812J：

```
Cisco1812J#ping 2001:db8:0:1::1
```

08-1 マルチベンダー機器によるOSPFv3基本設定

```
Type escape sequence to abort.
Sending 5, 100-byte ICMP Echos to 2001:DB8:0:1::1, timeout is 2 seconds:
!!!!!
Success rate is 100 percent (5/5), round-trip min/avg/max = 0/0/0 ms
Cisco1812J#ping 2001:db8:1:2::1

Type escape sequence to abort.
Sending 5, 100-byte ICMP Echos to 2001:DB8:1:2::1, timeout is 2 seconds:
!!!!!
Success rate is 100 percent (5/5), round-trip min/avg/max = 0/0/4 ms
```

■ AX620R1：

```
AX620R1(config)#ping6 2001:db8:0:3::1
PING 2001:db8:0:2::1 > 2001:db8:0:3::1 56 data bytes
64 bytes from 2001:db8:0:3::1 icmp_seq=0 hlim=64 time=1.180 ms
64 bytes from 2001:db8:0:3::1 icmp_seq=1 hlim=64 time=0.553 ms
64 bytes from 2001:db8:0:3::1 icmp_seq=2 hlim=64 time=0.567 ms
64 bytes from 2001:db8:0:3::1 icmp_seq=3 hlim=64 time=0.567 ms
64 bytes from 2001:db8:0:3::1 icmp_seq=4 hlim=64 time=0.561 ms

--- 2001:db8:0:3::1 ping statistics ---
5 packets transmitted, 5 packets received, 0% packet loss
round-trip (ms)  min/avg/max = 0.553/0.685/1.180
AX620R1(config)#ping6 2001:db8:1:2::1
PING 2001:db8:1:1::1 > 2001:db8:1:2::1 56 data bytes
64 bytes from 2001:db8:1:2::1 icmp_seq=0 hlim=64 time=0.982 ms
64 bytes from 2001:db8:1:2::1 icmp_seq=1 hlim=64 time=0.413 ms
64 bytes from 2001:db8:1:2::1 icmp_seq=2 hlim=64 time=0.403 ms
64 bytes from 2001:db8:1:2::1 icmp_seq=3 hlim=64 time=0.414 ms
64 bytes from 2001:db8:1:2::1 icmp_seq=4 hlim=64 time=0.415 ms

--- 2001:db8:1:2::1 ping statistics ---
5 packets transmitted, 5 packets received, 0% packet loss
round-trip (ms)  min/avg/max = 0.403/0.525/0.982
```

■ AX620R2:

```
AX620R2(config)#ping6 2001:db8:0:3::1
PING 2001:db8:2::2 > 2001:db8:0:3::1 56 data bytes
64 bytes from 2001:db8:0:3::1 icmp_seq=0 hlim=64 time=1.050 ms
64 bytes from 2001:db8:0:3::1 icmp_seq=1 hlim=64 time=0.567 ms
64 bytes from 2001:db8:0:3::1 icmp_seq=2 hlim=64 time=0.525 ms
64 bytes from 2001:db8:0:3::1 icmp_seq=3 hlim=64 time=0.528 ms
64 bytes from 2001:db8:0:3::1 icmp_seq=4 hlim=64 time=0.524 ms

--- 2001:db8:0:3::1 ping statistics ---
5 packets transmitted, 5 packets received, 0% packet loss
round-trip (ms)  min/avg/max = 0.524/0.638/1.050
AX620R2(config)#ping6 2001:db8:0:1::1
PING 2001:db8:1:1::2 > 2001:db8:0:1::1 56 data bytes
64 bytes from 2001:db8:0:1::1 icmp_seq=0 hlim=64 time=0.988 ms
64 bytes from 2001:db8:0:1::1 icmp_seq=1 hlim=64 time=0.415 ms
64 bytes from 2001:db8:0:1::1 icmp_seq=2 hlim=64 time=0.418 ms
64 bytes from 2001:db8:0:1::1 icmp_seq=3 hlim=64 time=0.406 ms
64 bytes from 2001:db8:0:1::1 icmp_seq=4 hlim=64 time=0.410 ms

--- 2001:db8:0:1::1 ping statistics ---
5 packets transmitted, 5 packets received, 0% packet loss
round-trip (ms)  min/avg/max = 0.406/0.527/0.988
```

08-2 シスコ機器によるOSPFv3基本設定

IPv6を用いたネットワーク上でOSPFv3を動作させる際に必要な基本的な項目について実習を行います。なお本実習では3台ともCiscoルーターを用います。

■実習トポロジー

この節で利用するトポロジーは以下のとおりです。

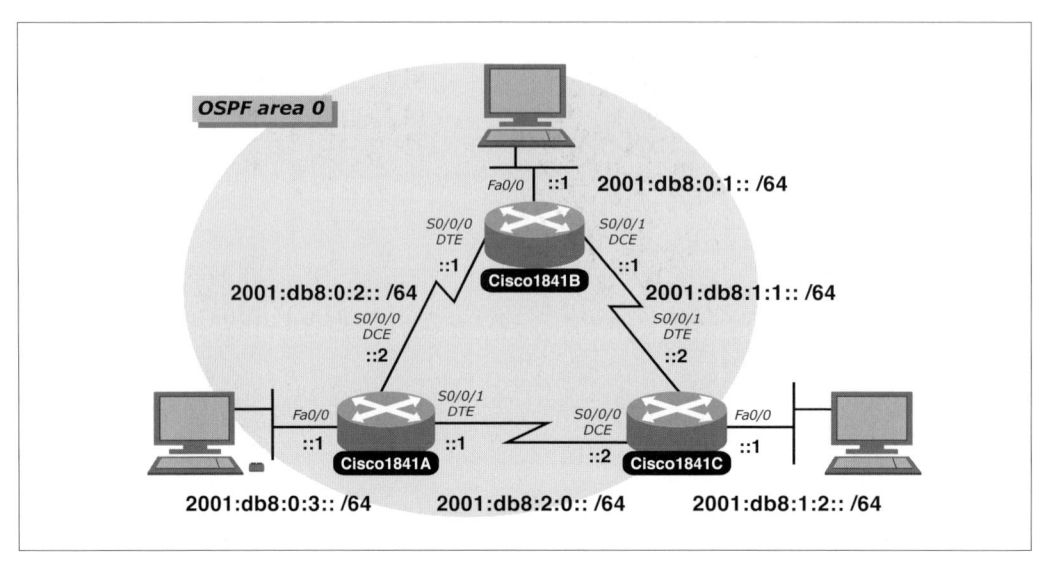

図8-2 ネットワークトポロジー

■インターフェース情報

表8-3 インターフェースアドレス割当一覧

ルーター	インターフェース	IPv6アドレス
Cisco1841A	FastEthernet 0/0 (Fa0/0)	2001:db8:0:3::1/64
Cisco1841A	Serial 0/0/0 (S0/0/0)	2001:db8:0:2::2/64
Cisco1841A	Serial 0/0/1 (S0/0/1)	2001:db8:2:0::1/64
Cisco1841B	FastEthernet 0/0 (Fa0/0)	2001:db8:0:1::1/64
Cisco1841B	Serial 0/0/0 (S0/0/0)	2001:db8:0:2::1/64
Cisco1841B	Serial 0/0/1 (S0/0/1)	2001:db8:1:1::1/64
Cisco1841C	FastEthernet 0/0 (Fa0/0)	2001:db8:1:2::1/64
Cisco1841C	Serial 0/0/0 (S0/0/0)	2001:db8:2:0::2/64
Cisco1841C	Serial 0/0/1 (S0/0/1)	2001:db8:1:1::2/64

■ルーターID情報

表8-4 ルーターID情報一覧

ルーター	ルーターID
Cisco1841A	10.0.0.1
Cisco1841B	172.16.1.1
Cisco1841C	192.168.1.1

■設定手順

STEP1 物理接続の実施

各ルーターおよびPCを"図8-2 ネットワークトポロジー"のとおりに接続します。DCE/DTEは図内表記を参照してください。

STEP2 ホスト名の設定とIPv6の有効化

CiscoルーターのIPv6トラフィック転送有効化とホスト名の設定を行います。

■ Cisco1841A：

```
Router(config)#hostname Cisco1841A
Cisco1841A(config)#ipv6 unicast-routing
```

■ Cisco1841B：

```
Router(config)#hostname Cisco1841B
Cisco1841B(config)#ipv6 unicast-routing
```

■ Cisco1841C：

```
Router(config)#hostname Cisco1841C
Cisco1841C(config)#ipv6 unicast-routing
```

STEP3 インターフェースへのIPv6アドレスの設定

すべてのルーターにIPv6アドレスを設定し，インターフェースを有効化します。アドレスは"表8-3 インターフェースアドレス割当一覧"を参照してください。

■ Cisco1841A：

```
Cisco1841A(config)#interface fastethernet0/0
Cisco1841A(config-if)#ipv6 enable
Cisco1841A(config-if)#ipv6 address 2001:db8:0:3::1/64
Cisco1841A(config-if)#no shutdown
Cisco1841A(config)#interface serial0/0/0
Cisco1841A(config-if)#ipv6 enable
Cisco1841A(config-if)#ipv6 address 2001:db8:0:2::2/64
Cisco1841A(config-if)#clock rate 64000
Cisco1841A(config-if)#no shutdown
Cisco1841A(config)#interface serial0/0/1
Cisco1841A(config-if)#ipv6 enable
Cisco1841A(config-if)#ipv6 address 2001:db8:2::1/64
Cisco1841A(config-if)#no shutdown
```

■ Cisco1841B：

```
Cisco1841B(config)#interface fastethernet0/0
Cisco1841B(config-if)#ipv6 enable
Cisco1841B(config-if)#ipv6 address 2001:db8:0:1::1/64
Cisco1841B(config-if)#no shutdown
Cisco1841B(config)#interface serial0/0/0
Cisco1841B(config-if)#ipv6 enable
Cisco1841B(config-if)#ipv6 address 2001:db8:0:2::1/64
Cisco1841B(config-if)#no shutdown
```

Chapter 08 OSPFv3の設定

```
Cisco1841B(config)#interface serial0/0/1
Cisco1841B(config-if)#ipv6 enable
Cisco1841B(config-if)#ipv6 address 2001:db8:1:1::1/64
Cisco1841B(config-if)#clock rate 64000
Cisco1841B(config-if)#no shutdown
```

■ Cisco1841C：

```
Cisco1841C(config)#interface fastethernet0/0
Cisco1841C(config-if)#ipv6 enable
Cisco1841C(config-if)#ipv6 address 2001:db8:1:2::1/64
Cisco1841C(config-if)#no shutdown
Cisco1841C(config)#interface serial0/0/0
Cisco1841C(config-if)#ipv6 enable
Cisco1841C(config-if)#ipv6 address 2001:db8:2::2/64
Cisco1841C(config-if)#clock rate 64000
Cisco1841C(config-if)#no shutdown
Cisco1841C(config)#interface serial0/0/1
Cisco1841C(config-if)#ipv6 enable
Cisco1841C(config-if)#ipv6 address 2001:db8:1:1::2/64
Cisco1841C(config-if)#no shutdown
```

STEP4 インターフェースのアドレス確認

インターフェースのステータスと動作していることを確認します。

`show ipv6 interface brief` コマンドを用いてインターフェース情報を表示しましょう。下記に出力を示します。インターフェース名の後の表示が [up/up] なら動作しています。

■ Cisco1841A：

`show ipv6 interface brief` コマンドで確認します。

```
Cisco1841A#show ipv6 interface brief
FastEthernet0/0            [up/up]
    FE80::21D:70FF:FEE6:33D0
    2001:DB8:0:3::1
FastEthernet0/1            [administratively down/down]
Serial0/0/0                [up/up]
    FE80::21D:70FF:FEE6:33D0
    2001:DB8:0:2::2
Serial0/0/1                [up/up]
    FE80::21D:70FF:FEE6:33D0
    2001:DB8:2::1
```

■ Cisco1841B：

`show ipv6 interface brief` コマンドで確認します。

```
Cisco1841B#show ipv6 interface brief
FastEthernet0/0            [up/up]
    FE80::211:21FF:FE67:B480
    2001:DB8:0:1::1
Serial0/0/0                [up/up]
    FE80::211:21FF:FE67:B480
    2001:DB8:0:2::1
FastEthernet0/1            [administratively down/down]
Serial0/0/1                [up/up]
    FE80::211:21FF:FE67:B480
    2001:DB8:1:1::1
```

■ Cisco1841C:

`show ipv6 interface brief` コマンドで確認します。

```
Cisco1841C#show ipv6 interface brief
FastEthernet0/0            [up/up]
    FE80::21D:70FF:FEE6:3268
    2001:DB8:1:2::1
FastEthernet0/1            [administratively down/down]
Serial0/0/0                [up/up]
    FE80::21D:70FF:FEE6:3268
    2001:DB8:2::2
Serial0/0/1                [up/up]
    FE80::21D:70FF:FEE6:3268
    2001:DB8:1:1::2
```

STEP5 OSPFv3 プロセスの設定

OSPFv3 をプロセス番号 1 で起動し，各ルーターにルーター ID を設定します。ルーター ID は " 表 8-4　ルーター ID 情報一覧 " を参照してください。

■ Cisco1841A:

```
Cisco1841A(config)#ipv6 router ospf 1
Cisco1841A(config-router)#router-id 10.0.0.1
```

■ Cisco1841B:

```
Cisco1841B(config)#ipv6 router ospf 1
Cisco1841B(config-router)#router-id 172.16.1.1
```

■ Cisco1841C:

```
Cisco1841C(config)#ipv6 router ospf 1
Cisco1841C(config-router)#router-id 192.168.1.1
```

STEP6 インターフェースへの OSPFv3 プロセス及びエリアの割当

各インターフェースに OSPFv3 プロセス 1，エリア 0 を割り当てます。

■ Cisco1841A:

```
Cisco1841A(config)#interface fastethernet0/0
Cisco1841A(config-if)#ipv6 ospf 1 area 0
Cisco1841A(config)#interface serial0/0/0
Cisco1841A(config-if)#ipv6 ospf 1 area 0
Cisco1841A(config)#interface serial0/0/1
Cisco1841A(config-if)#ipv6 ospf 1 area 0
```

■ Cisco1841B:

```
Cisco1841B(config)#interface fastethernet0/0
Cisco1841B(config-if)#ipv6 ospf 1 area 0
Cisco1841B(config)#interface serial0/0/0
Cisco1841B(config-if)#ipv6 ospf 1 area 0
Cisco1841B(config)#interface serial0/0/1
Cisco1841B(config-if)#ipv6 ospf 1 area 0
```

Chapter 08 OSPFv3の設定

■ Cisco1841C:

```
Cisco1841C(config)#interface fastethernet0/0
Cisco1841C(config-if)#ipv6 ospf 1 area 0
Cisco1841C(config)#interface serial0/0/0
Cisco1841C(config-if)#ipv6 ospf 1 area 0
Cisco1841C(config)#interface serial0/0/1
Cisco1841C(config-if)# ipv6 ospf 1 area 0
```

STEP7 OSPFv3 設定確認

各ルーターで動作しているルーティングプロトコルが設定した内容 (OSPFv3 process1 Area0) であることを **show ipv6 protocols** コマンドで確認します。

■ Cisco1841A:

show ipv6 protocols コマンドで確認します。

```
Cisco1841A#show ipv6 protocols
IPv6 Routing Protocol is "connected"
IPv6 Routing Protocol is "static"
IPv6 Routing Protocol is "ospf 1"
  Interfaces (Area 0):
    Serial0/0/1
    Serial0/0/0
    FastEthernet0/0
  Redistribution:
    None
```

■ Cisco1841B:

show ipv6 protocols コマンドで確認します。

```
Cisco1841B#show ipv6 protocols
IPv6 Routing Protocol is "connected"
IPv6 Routing Protocol is "static"
IPv6 Routing Protocol is "ospf 1"
  Interfaces (Area 0):
    Serial0/0/1
    Serial0/0/0
    FastEthernet0/0
  Redistribution:
    None
```

■ Cisco1841C:

show ipv6 protocols コマンドで確認します。

```
Cisco1841C#show ipv6 protocols
IPv6 Routing Protocol is "connected"
IPv6 Routing Protocol is "static"
IPv6 Routing Protocol is "ospf 1"
  Interfaces (Area 0):
    Serial0/0/1
    Serial0/0/0
    FastEthernet0/0
  Redistribution:
    None
```

STEP8　ルーター相互の隣接情報確認

各ルーターの隣接情報などを **show ipv6 ospf neighbor** コマンドで確認します。6-1節の実習と異なり，シリアルを用いた接続のため DR, BDR の選出が行われないことに注意してください。

■ Cisco1841A：

show ipv6 ospf neighbor コマンドで確認します。

```
Cisco1841A#show ipv6 ospf neighbor

Neighbor ID     Pri   State          Dead Time   Interface ID    Interface
192.168.1.1       1   FULL/  -       00:00:37    6               Serial0/0/1
172.16.1.1        1   FULL/  -       00:00:32    6               Serial0/0/0
```

■ Cisco1841B：

show ipv6 ospf neighbor コマンドで確認します。

```
Cisco1841B#show ipv6 ospf neighbor

Neighbor ID     Pri   State          Dead Time   Interface ID    Interface
192.168.1.1       1   FULL/  -       00:00:33    7               Serial0/0/1
10.0.0.1          1   FULL/  -       00:00:38    6               Serial0/0/0
```

■ Cisco1841C：

show ipv6 ospf neighbor コマンドで確認します。

```
Cisco1841C#show ipv6 ospf neighbor

Neighbor ID     Pri   State          Dead Time   Interface ID    Interface
172.16.1.1        1   FULL/  -       00:00:35    7               Serial0/0/1
10.0.0.1          1   FULL/  -       00:00:38    7               Serial0/0/0
```

STEP9　OSPFv3 データベースの確認

さらに各ルーターの OSPFv3 データベースの内容を **show ipv6 ospf database** コマンドで確認します。

■ Cisco1841A：

show ipv6 ospf database コマンドで確認します。

```
Cisco1841A#show ipv6 ospf database

            OSPFv3 Router with ID (10.0.0.1) (Process ID 1)

                Router Link States (Area 0)

ADV Router         Age         Seq#           Fragment ID   Link count   Bits
10.0.0.1           1172        0x80000005     0             2            None
172.16.1.1         1457        0x80000003     0             2            None
192.168.1.1        1194        0x80000005     0             2            None

                Link (Type-8) Link States (Area 0)

ADV Router         Age         Seq#           Link ID       Interface
10.0.0.1           1172        0x80000003     7             Se0/0/1
192.168.1.1        1194        0x80000003     6             Se0/0/1
```

Chapter 08 OSPFv3の設定

```
           10.0.0.1              1416          0x80000002      6          Se0/0/0
           172.16.1.1            1457          0x80000002      6          Se0/0/0
           10.0.0.1              1416          0x80000002      4          Fa0/0

                        Intra Area Prefix Link States (Area 0)

           ADV Router            Age           Seq#            Link ID    Ref-lstype    Ref-LSID
           10.0.0.1              1172          0x80000003      0          0x2001        0
           172.16.1.1            1457          0x80000002      0          0x2001        0
           192.168.1.1           1201          0x80000004      0          0x2001        0
```

■ **Cisco1841B：**

show ipv6 ospf database コマンドで確認します。

```
Cisco1841B#show ipv6 ospf database

            OSPFv3 Router with ID (172.16.1.1) (Process ID 1)

            Router Link States (Area 0)

ADV Router            Age           Seq#            Fragment ID   Link count    Bits
10.0.0.1              1342          0x80000005      0             2             None
172.16.1.1            1625          0x80000003      0             2             None
192.168.1.1           1362          0x80000005      0             2             None

            Link (Type-8) Link States (Area 0)

ADV Router            Age           Seq#            Link ID       Interface
172.16.1.1            1625          0x80000002      7             Se0/0/1
192.168.1.1           1611          0x80000003      7             Se0/0/1
10.0.0.1              1586          0x80000002      6             Se0/0/0
172.16.1.1            1625          0x80000002      6             Se0/0/0
172.16.1.1            1625          0x80000002      4             Fa0/0

            Intra Area Prefix Link States (Area 0)

ADV Router            Age           Seq#            Link ID       Ref-lstype    Ref-LSID
10.0.0.1              1342          0x80000003      0             0x2001        0
172.16.1.1            1625          0x80000002      0             0x2001        0
192.168.1.1           1364          0x80000004      0             0x2001        0
```

■ **Cisco1841C：**

show ipv6 ospf database コマンドで確認します。

```
Cisco1841C#show ipv6 ospf database

            OSPFv3 Router with ID (192.168.1.1) (Process ID 1)

            Router Link States (Area 0)

ADV Router            Age           Seq#            Fragment ID   Link count    Bits
10.0.0.1              1386          0x80000005      0             2             None
172.16.1.1            1670          0x80000003      0             2             None
192.168.1.1           1406          0x80000005      0             2             None

            Link (Type-8) Link States (Area 0)

ADV Router            Age           Seq#            Link ID       Interface
172.16.1.1            1670          0x80000002      7             Se0/0/1
```

```
192.168.1.1         1655        0x80000003      7           Se0/0/1
10.0.0.1            1386        0x80000003      7           Se0/0/0
192.168.1.1         1406        0x80000003      6           Se0/0/0
192.168.1.1         1655        0x80000002      4           Fa0/0


            Intra Area Prefix Link States (Area 0)

ADV Router          Age         Seq#            Link ID     Ref-lstype      Ref-LSID
10.0.0.1            1386        0x80000003      0           0x2001          0
172.16.1.1          1670        0x80000002      0           0x2001          0
192.168.1.1         1408        0x80000004      0           0x2001          0
```

> **STEP10** ルーティングテーブルの確認

 Show ipv6 route コマンドを用いて，ルーティングテーブル情報を表示します。下記に出力例を示します。先頭にOの表記があるものはOSPFv3により生成した経路です。

■ Cisco1841A：

show ipv6 route コマンドで確認します。

```
Cisco1841A#show ipv6 route
IPv6 Routing Table - 11 entries
Codes: C - Connected, L - Local, S - Static, R - RIP, B - BGP
       U - Per-user Static route
       I1 - ISIS L1, I2 - ISIS L2, IA - ISIS interarea, IS - ISIS summary
       O - OSPF intra, OI - OSPF inter, OE1 - OSPF ext 1, OE2 - OSPF ext 2
       ON1 - OSPF NSSA ext 1, ON2 - OSPF NSSA ext 2
O   2001:DB8:0:1::/64 [110/782]
     via FE80::211:21FF:FE67:B480, Serial0/0/0
C   2001:DB8:0:2::/64 [0/0]
     via ::, Serial0/0/0
L   2001:DB8:0:2::2/128 [0/0]
     via ::, Serial0/0/0
C   2001:DB8:0:3::/64 [0/0]
     via ::, FastEthernet0/0
L   2001:DB8:0:3::1/128 [0/0]
     via ::, FastEthernet0/0
O   2001:DB8:1:1::/64 [110/1562]
     via FE80::21D:70FF:FEE6:3268, Serial0/0/1
     via FE80::211:21FF:FE67:B480, Serial0/0/0
O   2001:DB8:1:2::/64 [110/782]
     via FE80::21D:70FF:FEE6:3268, Serial0/0/1
C   2001:DB8:2::/64 [0/0]
     via ::, Serial0/0/1
L   2001:DB8:2::1/128 [0/0]
     via ::, Serial0/0/1
L   FE80::/10 [0/0]
     via ::, Null0
L   FF00::/8 [0/0]
     via ::, Null0
```

■ Cisco1841B：

show ipv6 route コマンドで確認します。

```
Cisco1841B#show ipv6 route
IPv6 Routing Table - 11 entries
Codes: C - Connected, L - Local, S - Static, R - RIP, B - BGP
```

Chapter 08 OSPFv3の設定

```
               U - Per-user Static route
               I1 - ISIS L1, I2 - ISIS L2, IA - ISIS interarea, IS - ISIS summary
               O - OSPF intra, OI - OSPF inter, OE1 - OSPF ext 1, OE2 - OSPF ext 2
               ON1 - OSPF NSSA ext 1, ON2 - OSPF NSSA ext 2
C   2001:DB8:0:1::/64 [0/0]
     via ::, FastEthernet0/0
L   2001:DB8:0:1::1/128 [0/0]
     via ::, FastEthernet0/0
C   2001:DB8:0:2::/64 [0/0]
     via ::, Serial0/0/0
L   2001:DB8:0:2::1/128 [0/0]
     via ::, Serial0/0/0
O   2001:DB8:0:3::/64 [110/782]
     via FE80::21D:70FF:FEE6:33D0, Serial0/0
C   2001:DB8:1:1::/64 [0/0]
     via ::, Serial0/0/1
L   2001:DB8:1:1::1/128 [0/0]
     via ::, Serial0/0/1
O   2001:DB8:1:2::/64 [110/782]
     via FE80::21D:70FF:FEE6:3268, Serial0/0/1
O   2001:DB8:2::/64 [110/1562]
     via FE80::21D:70FF:FEE6:3268, Serial0/0/1
     via FE80::21D:70FF:FEE6:33D0, Serial0/0/0
L   FE80::/10 [0/0]
     via ::, Null0
L   FF00::/8 [0/0]
     via ::, Null0
```

■ **Cisco1841C：**

show ipv6 route コマンドで確認します。

```
Cisco1841C#show ipv6 route
IPv6 Routing Table - 11 entries
Codes: C - Connected, L - Local, S - Static, R - RIP, B - BGP
       U - Per-user Static route
       I1 - ISIS L1, I2 - ISIS L2, IA - ISIS interarea, IS - ISIS summary
       O - OSPF intra, OI - OSPF inter, OE1 - OSPF ext 1, OE2 - OSPF ext 2
       ON1 - OSPF NSSA ext 1, ON2 - OSPF NSSA ext 2
O   2001:DB8:0:1::/64 [110/782]
     via FE80::211:21FF:FE67:B480, Serial0/0/1
O   2001:DB8:0:2::/64 [110/1562]
     via FE80::21D:70FF:FEE6:33D0, Serial0/0/0
     via FE80::211:21FF:FE67:B480, Serial0/0/1
O   2001:DB8:0:3::/64 [110/782]
     via FE80::21D:70FF:FEE6:33D0, Serial0/0/0
C   2001:DB8:1:1::/64 [0/0]
     via ::, Serial0/0/1
L   2001:DB8:1:1::2/128 [0/0]
     via ::, Serial0/0/1
C   2001:DB8:1:2::/64 [0/0]
     via ::, FastEthernet0/0
L   2001:DB8:1:2::1/128 [0/0]
     via ::, FastEthernet0/0
C   2001:DB8:2::/64 [0/0]
     via ::, Serial0/0/0
L   2001:DB8:2::2/128 [0/0]
     via ::, Serial0/0/0
L   FE80::/10 [0/0]
     via ::, Null0
```

```
L   FF00::/8 [0/0]
     via ::, Null0
Cisco1841C#
```

STEP11 通信による確認

各ルーターから直接つながっていない別のルーターのインターフェースまでの通信を **ping** コマンドで確認します。

■ Cisco1841A：

```
Cisco1841A#ping 2001:db8:0:1::1

Type escape sequence to abort.
Sending 5, 100-byte ICMP Echos to 2001:DB8:0:1::1, timeout is 2 seconds:
!!!!!
Success rate is 100 percent (5/5), round-trip min/avg/max = 28/28/28 ms
Cisco1841A#ping 2001:db8:1:2::1

Type escape sequence to abort.
Sending 5, 100-byte ICMP Echos to 2001:DB8:1:2::1, timeout is 2 seconds:
!!!!!
Success rate is 100 percent (5/5), round-trip min/avg/max = 28/28/28 ms
```

■ Cisco1841B：

```
Cisco1841B#ping 2001:db8:0:3::1

Type escape sequence to abort.
Sending 5, 100-byte ICMP Echos to 2001:DB8:0:3::1, timeout is 2 seconds:
!!!!!
Success rate is 100 percent (5/5), round-trip min/avg/max = 28/28/32 ms
Cisco1841B#ping 2001:db8:1:2::1

Type escape sequence to abort.
Sending 5, 100-byte ICMP Echos to 2001:DB8:1:2::1, timeout is 2 seconds:
!!!!!
Success rate is 100 percent (5/5), round-trip min/avg/max = 28/28/28 ms
```

■ Cisco1841C：

```
Cisco1841C#ping 2001:db8:0:1::1

Type escape sequence to abort.
Sending 5, 100-byte ICMP Echos to 2001:DB8:0:1::1, timeout is 2 seconds:
!!!!!
Success rate is 100 percent (5/5), round-trip min/avg/max = 28/28/28 ms
Cisco1841C#ping 2001:db8:0:3::1

Type escape sequence to abort.
Sending 5, 100-byte ICMP Echos to 2001:DB8:0:3::1, timeout is 2 seconds:
!!!!!
Success rate is 100 percent (5/5), round-trip min/avg/max = 28/28/28 ms
```

Chapter 09

OSPFv3の デフォルトルートの伝搬

09-1 マルチベンダー機器による OSPFv3のデフォルトルートの伝搬

09-2 シスコ機器によるOSPFv3 のデフォルトルートの伝搬

09-1 マルチベンダー機器によるOSPFv3のデフォルトルートの伝搬

IPv6を用いたネットワーク上でOSPFv3を動作させ，デフォルトルートを伝搬させる実習です。なお本実習では1台のCiscoルーター (Cisco1812J) と2台のAX620Rを用います。

■実習トポロジー

AX620R2のFastEternet 0/0.0をデフォルトルートとします。この情報をOSPFv3へ伝搬させます。

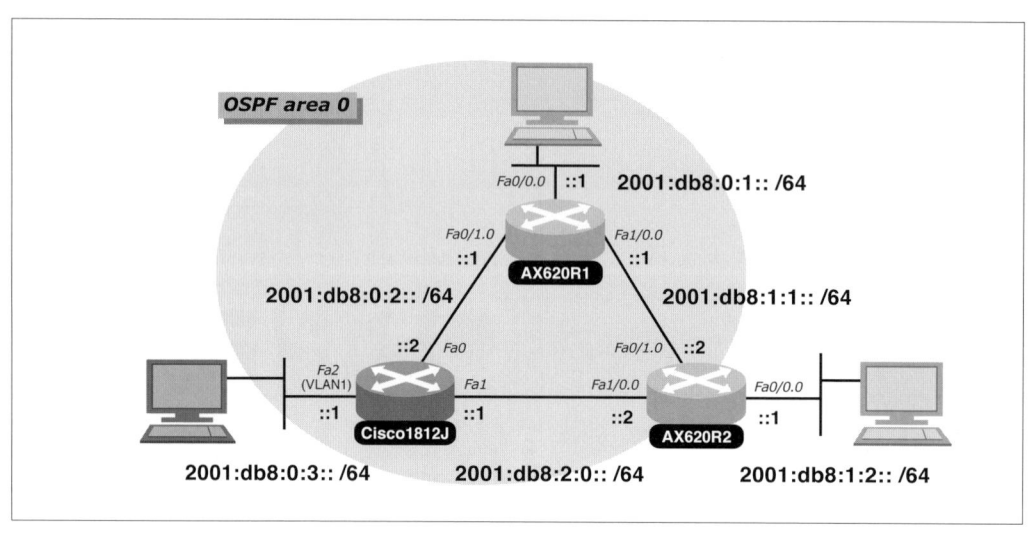

図9-1　ネットワークトポロジー

■インターフェース情報

表9-1　インターフェースアドレス割当一覧

ルーター	インターフェース	IPv6アドレス
Cisco1812J	FastEthernet 0 (Fa0)	2001:db8:0:2::2/64
	FastEthernet 1 (Fa1)	2001:db8:2:0::1/64
	FastEthernet 2 (VLAN1) (Fa2)	2001:db8:0:3::1/64
AX620R1	FastEthernet 0/0.0 (Fa0/0.0)	2001:db8:0:1::1/64
	FastEthernet 0/1.0 (Fa0/1.0)	2001:db8:0:2::1/64
	FastEthernet 1/0.0 (Fa1/0.0)	2001:db8:1:1::1/64
AX620R2	FastEthernet 0/0.0 (Fa0/0.0)	2001:db8:1:2::1/64
	FastEthernet 0/1.0 (Fa0/1.0)	2001:db8:1:1::2/64
	FastEthernet 1/0.0 (Fa1/0.0)	2001:db8:2:0::2/64

■ルーターID情報

表9-2　ルーターID情報一覧

ルーター	ルーターID
Cisco1812J	10.0.0.1
AX620R1	172.16.1.1
AX620R2	192.168.1.1

■設定手順

STEP1 物理接続の実施

各ルーターおよび PC を "図 9-1　ネットワークトポロジー" のとおりに接続します。

STEP2 ホスト名の設定と IPv6 の有効化

ルーターのホスト名と Cisco ルーターに IPv6 トラフィック転送有効化の設定を行います。

■ Cisco1812J：

```
Router(config)#hostname Cisco1812J
Cisco1812J(config)#ipv6 unicast-routing
```

■ AX620R1：

```
Router(config)#hostname AX620R1
```

■ AX620R2：

```
Router(config)#hostname AX620R2
```

STEP3 インターフェースへの IPv6 アドレスの設定

すべてのルーターに IPv6 アドレスを設定し，インターフェースを有効化します。アドレスは "表 9-1　インターフェースアドレス割当一覧" を参照してください。なお，Cisco1812J は FastEthernet 2 から 6 までスイッチポートとして動作するため，VLAN1 として設定します。

■ Cisco1812J：

```
Cisco1812J(config)#interface fastethernet0
Cisco1812J(config-if)#ipv6 enable
Cisco1812J(config-if)#ipv6 address 2001:db8:0:2::2/64
Cisco1812J(config-if)#no shutdown
Cisco1812J(config)#interface FastEthernet1
Cisco1812J(config-if)#ipv6 enable
Cisco1812J(config-if)#ipv6 address 2001:db8:2:0::1/64
Cisco1812J(config-if)#no shutdown
Cisco1812J(config)#interface vlan 1
Cisco1812J(config-if)#ipv6 enable
Cisco1812J(config-if)#ipv6 address 2001:db8:0:3::1/64
Cisco1812J(config-if)#no shutdown
```

■ AX620R1：

```
AX620R1(config)#interface FastEthernet0/0.0
AX620R1(config-FastEthernet0/0.0)#ipv6 enable
AX620R1(config-FastEthernet0/0.0)#ipv6 address 2001:db8:0:1::1/64
AX620R1(config-FastEthernet0/0.0)#no shutdown
AX620R1(config)#interface FastEthernet0/1.0
AX620R1(config-FastEthernet0/1.0)#ipv6 enable
AX620R1(config-FastEthernet0/1.0)#ipv6 address 2001:db8:0:2::1/64
AX620R1(config-FastEthernet0/1.0)#no shutdown
AX620R1(config)#interface FastEthernet1/0.0
AX620R1(config-FastEthernet1/0.0)#ipv6 enable
AX620R1(config-FastEthernet1/0.0)#ipv6 address 2001:db8:1:1::1/64
```

Chapter 09 OSPFv3のデフォルトルートの伝搬

```
AX620R1(config-FastEthernet1/0.0)#no shutdown
```

■ AX620R2：

```
AX620R2(config)#interface FastEthernet0/0.0
AX620R2(config-FastEthernet0/0.0)#ipv6 enable
AX620R2(config-FastEthernet0/0.0)#ipv6 address 2001:db8:1:2::1/64
AX620R2(config-FastEthernet0/0.0)#no shutdown
AX620R2(config)#interface FastEthernet1/0.0
AX620R2(config-FastEthernet1/0.0)#ipv6 enable
AX620R2(config-FastEthernet1/0.0)#ipv6 address 2001:db8:2:0::2/64
AX620R2(config-FastEthernet1/0.0)#no shutdown
AX620R2(config)#interface FastEthernet0/1.0
AX620R2(config-FastEthernet0/1.0)#ipv6 enable
AX620R2(config-FastEthernet0/1.0)#ipv6 address 2001:db8:1:1::2/64
AX620R2(config-FastEthernet0/1.0)#no shutdown
```

STEP4 インターフェースのアドレス確認

インターフェースのステータスと動作していることを確認します。

`show ipv6 interface brief` または `show ipv6 address` コマンドを用いてインターフェース情報を表示させます。下記に出力を示します。インターフェース名の後の表示が [up/up] なら動作しています。

■ Cisco1812J：

`show ipv6 interface brief` コマンドで確認します。

```
Cisco1812J#show ipv6 interface brief
FastEthernet0              [up/up]
    FE80::226:BFF:FEF6:B5FA
    2001:DB8:0:2::2
FastEthernet1              [up/up]
    FE80::226:BFF:FEF6:B5FB
    2001:DB8:2::1
BRI0                       [administratively down/down]
BRI0:1                     [administratively down/down]
BRI0:2                     [administratively down/down]
FastEthernet2              [up/up]
FastEthernet3              [up/down]
FastEthernet4              [up/down]
FastEthernet5              [up/down]
FastEthernet6              [up/down]
FastEthernet7              [up/down]
FastEthernet8              [up/down]
FastEthernet9              [down/down]
Vlan1                      [up/up]
    FE80::226:BFF:FEF6:B5FA
    2001:DB8:0:3::1
```

■ AX620R1：

`show ipv6 address` コマンドで確認します。

```
AX620R1(config)#show ipv6 address
Interface FastEthernet0/0.0 is up, line protocol is up
  Global address(es):
    2001:db8:0:1::1 prefixlen 64
    2001:db8:0:1::0 prefixlen 64 anycast
```

09-1 マルチベンダー機器によるOSPFv3のデフォルトルートの伝搬

```
    Link-local address(es):
      fe80::260:b9ff:fe4b:a477 prefixlen 64
      fe80::0 prefixlen 64 anycast
    Multicast address(es):
      ff02::1
      ff02::2
      ff02::5
      ff02::6
      ff02::1:ff00:0
      ff02::1:ff00:1
      ff02::1:ff4b:a477
Interface FastEthernet0/1.0 is up, line protocol is up
    Global address(es):
      2001:db8:0:2::1 prefixlen 64
      2001:db8:0:2::0 prefixlen 64 anycast
    Link-local address(es):
      fe80::260:b9ff:fe4b:a4f7 prefixlen 64
      fe80::0 prefixlen 64 anycast
    Multicast address(es):
      ff02::1
      ff02::2
      ff02::5
      ff02::6
      ff02::1:ff00:0
      ff02::1:ff00:1
      ff02::1:ff4b:a4f7
Interface FastEthernet1/0.0 is up, line protocol is up
    Global address(es):
      2001:db8:1:1::1 prefixlen 64
      2001:db8:1:1::0 prefixlen 64 anycast
    Link-local address(es):
      fe80::260:b9ff:fe4b:a40f prefixlen 64
      fe80::0 prefixlen 64 anycast
    Multicast address(es):
      ff02::1
      ff02::2
      ff02::5
      ff02::6
      ff02::1:ff00:0
      ff02::1:ff00:1
      ff02::1:ff4b:a40f
Interface Loopback0.0 is up, line protocol is up
    Orphan address(es):
      ::1 prefixlen 128
Interface Loopback1.0 is up, line protocol is up
Interface Null0.0 is up, line protocol is up
Interface Null1.0 is up, line protocol is up
```

■ AX620R2：

show ipv6 address コマンドで確認します。

```
AX620R2(config)#show ipv6 address
Interface FastEthernet0/0.0 is up, line protocol is up
    Global address(es):
      2001:db8:1:2::1 prefixlen 64
      2001:db8:1:2:: prefixlen 64 anycast
    Link-local address(es):
      fe80::260:b9ff:fe4f:225f prefixlen 64
      fe80:: prefixlen 64 anycast
```

Chapter 09 OSPFv3のデフォルトルートの伝搬

```
    Multicast address(es):
      ff02::1
      ff02::2
      ff02::5
      ff02::6
      ff02::1:ff00:0
      ff02::1:ff00:1
      ff02::1:ff4f:225f
  Interface FastEthernet0/1.0 is up, line protocol is up
    Global address(es):
      2001:db8:1:1::2 prefixlen 64
      2001:db8:1:1:: prefixlen 64 anycast
    Link-local address(es):
      fe80::260:b9ff:fe4f:22df prefixlen 64
      fe80:: prefixlen 64 anycast
    Multicast address(es):
      ff02::1
      ff02::2
      ff02::5
      ff02::6
      ff02::1:ff00:0
      ff02::1:ff00:2
      ff02::1:ff4f:22df
  Interface FastEthernet1/0.0 is up, line protocol is up
    Global address(es):
      2001:db8:2::2 prefixlen 64
      2001:db8:2:: prefixlen 64 anycast
    Link-local address(es):
      fe80::260:b9ff:fe4f:223f prefixlen 64
      fe80:: prefixlen 64 anycast
    Multicast address(es):
      ff02::1
      ff02::2
      ff02::5
      ff02::6
      ff02::1:ff00:0
      ff02::1:ff00:2
      ff02::1:ff4f:223f
  Interface Loopback0.0 is up, line protocol is up
    Orphan address(es):
      ::1 prefixlen 128
  Interface Loopback1.0 is up, line protocol is up
  Interface Null0.0 is up, line protocol is up
  Interface Null1.0 is up, line protocol is up
```

STEP5　OSPFv3プロセスの設定

　OSPFv3をプロセス番号1で起動し、各ルーターにルーターIDを設定します。ルーターIDは"表9-2　ルーターID情報一覧"を参照してください。

■ Cisco1812J：

```
Cisco1812J(config)#ipv6 router ospf 1
Cisco1812J(config-router)#router-id 10.0.0.1
```

■ AX620R1：

```
AX620R1(config)#ipv6 router ospf 1
AX620R1(config-ospfv3-1)#router-id 172.16.1.1
```

■ AX620R2：

```
AX620R2(config)#ipv6 router ospf 1
AX620R2(config-ospfv3-1)#router-id 192.168.1.1
```

STEP6　インターフェースへの OSPFv3 プロセス及びエリアの割当

各インターフェースに OSPFv3 プロセス 1，エリア 0 を割り当てます。

■ Cisco1812J：

```
Cisco1812J(config)#interface FastEthernet0
Cisco1812J(config-if)#ipv6 ospf 1 area 0
Cisco1812J(config)#interface FastEthernet1
Cisco1812J(config-if)#ipv6 ospf 1 area 0
Cisco1812J(config)#interface vlan1
Cisco1812J(config-if)#ipv6 ospf 1 area 0
```

■ AX620R1：

```
AX620R1(config-ospfv3-1)#network FastEthernet0/0.0 area 0
AX620R1(config-ospfv3-1)#network FastEthernet0/1.0 area 0
AX620R1(config-ospfv3-1)#network FastEthernet1/0.0 area 0
```

■ AX620R2：

```
AX620R2(config-ospfv3-1)#network FastEthernet0/1.0 area 0
AX620R2(config-ospfv3-1)#network FastEthernet1/0.0 area 0
```

STEP7　デフォルトルートのインターフェースへの設定

AX620R2 ルーター上でデフォルトルートを FastEthernet 0/0.0 に設定します。

■ AX620R2：

```
AX620R2(config)#ipv6 route ::/0 FastEthernet0/0.0
```

STEP8　デフォルトルートの OSPFv3 への伝搬

AX620R2 ルーター上でデフォルトルートをアドバタイズさせます。

```
AX620R2(config)#ipv6 router ospf 1
AX620R2(config-ospfv3-1)#originate-default
```

STEP9　OSPFv3 設定確認

各ルーターで動作しているルーティングプロトコルが設定した内容 (OSPFv3 process1 Area0) であることを **show ipv6 protocols** コマンドで確認します。

■ Cisco1812J：

show ipv6 protocols コマンドで確認します。

```
Cisco1812J#show ipv6 protocols
IPv6 Routing Protocol is "connected"
IPv6 Routing Protocol is "static"
IPv6 Routing Protocol is "ospf 1"
```

```
    Interfaces (Area 0):
      Vlan1
      FastEthernet1
      FastEthernet0
    Redistribution:
      None
```

■ **AX620R1**：

show ipv6 protocols コマンドで確認します。

```
AX620R1(config)#show ipv6 protocols
IPv6 unicast routing is enabled
  Process switching queue-len: 1/3 (last/peak), overflows: 0
  Host transit queue-len: 3/5 (last/peak), overflows: 0
  Fragment transit queue-len: 0/0 (last/peak), overflows: 0
  FIB entries:
    System: 40/unlimited (busy/max)
    Dynamic routing: 7/unlimited (busy/max)
  Routing cache bucket(s): 0/4096/0/4096 (busy/free/garbage/max)
  Maximum path(s): 16
  Load sharing algorithm is per-packet round robin
IPv6 multicast routing is disabled
IPv6 source-routing type-0 is disabled
  Received 0 type-0, 0 other types
  Discarded 0 type-0, 0 other types
    0 truncated, 0 invalid address, 0 other error
IPv6 unified forwarding service cache is disabled
IPv6 reassembly service is enabled
  Reassemble buffer size is 0/65535 octets (peak/max)
  Reassemble buffers: 0/0/16 (curr/peak/max)
ICMP for IPv6 is enabled
  Error message limited to one every 1000 milliseconds
Neighbor discovery for IPv6 is enabled
  Neighbor cache(s): 1/1024 (busy/max)
Path MTU discovery is enabled
  Cache entries: 0/0/unlimited (curr/peak/max)
  System errors: 0/0 (overflows/alloc fails)
  Too big received: 0/0/0 (received/under 1280/over 65535)
IPv6 routing protocol is "ospf 1"
  Router ID 172.16.1.1
  RIB entries: 7
  Autonomous system boundary capability: no
  Area border capability: no
  Number of areas in this router: 1
  Originating default route is disabled
  Administrative distance:
    External 110, Intra-Area 110, Inter-Area 110
  Delay time between receiving a change to SPF calculations: 5 seconds
  Hold time between consecutive SPF calculations: 10 seconds
  Redistribution:
    None
  Passive interface(s):
    None
  Network(s):
    FastEthernet0/0.0, area 0(0.0.0.0)
    FastEthernet0/1.0, area 0(0.0.0.0)
    FastEthernet1/0.0, area 0(0.0.0.0)
```

■AX620R2：

show ipv6 protocols コマンドで確認します。

```
AX620R2(config)#show ipv6 protocols
IPv6 unicast routing is enabled
  Process switching queue-len: 1/3 (last/peak), overflows: 0
  Host transit queue-len: 3/4 (last/peak), overflows: 0
  Fragment transit queue-len: 0/0 (last/peak), overflows: 0
  FIB entries:
    System: 41/unlimited (busy/max)
    Dynamic routing: 8/unlimited (busy/max)
  Routing cache bucket(s): 0/4096/0/4096 (busy/free/garbage/max)
  Maximum path(s): 16
  Load sharing algorithm is per-packet round robin
IPv6 multicast routing is disabled
IPv6 source-routing type-0 is disabled
  Received 0 type-0, 0 other types
  Discarded 0 type-0, 0 other types
    0 truncated, 0 invalid address, 0 other error
IPv6 unified forwarding service cache is disabled
IPv6 reassembly service is enabled
  Reassemble buffer size is 0/65535 octets (peak/max)
  Reassemble buffers: 0/0/16 (curr/peak/max)
ICMP for IPv6 is enabled
  Error message limited to one every 1000 milliseconds
Neighbor discovery for IPv6 is enabled
  Neighbor cache(s): 0/1024 (busy/max)
Path MTU discovery is enabled
  Cache entries: 0/0/unlimited (curr/peak/max)
  System errors: 0/0 (overflows/alloc fails)
  Too big received: 0/0/0 (received/under 1280/over 65535)
IPv6 routing protocol is "ospf 1"
  Router ID 192.168.1.1
  RIB entries: 8
  Autonomous system boundary capability: yes
  Area border capability: no
  Number of areas in this router: 1
  Originating default route is enabled
  Administrative distance:
    External 110, Intra-Area 110, Inter-Area 110
  Delay time between receiving a change to SPF calculations: 5 seconds
  Hold time between consecutive SPF calculations: 10 seconds
  Redistribution:
    None
  Passive interface(s):
    None
  Network(s):
    FastEthernet0/0.0, area 0(0.0.0.0)
    FastEthernet0/1.0, area 0(0.0.0.0)
    FastEthernet1/0.0, area 0(0.0.0.0)
```

STEP10　ルーター相互の隣接情報確認

各ルーターの隣接情報などを **show ipv6 ospf neighbor** コマンドで確認します。

■Cisco1812J：

show ipv6 ospf neighbor コマンドで確認します。

Chapter 09 OSPFv3のデフォルトルートの伝搬

```
Cisco1812J#show ipv6 ospf neighbor

Neighbor ID     Pri   State           Dead Time   Interface ID    Interface
192.168.1.1      1    FULL/BDR        00:00:34        3           FastEthernet1
172.16.1.1       1    FULL/BDR        00:00:31        2           FastEthernet0
```

■ AX620R1：

show ipv6 ospf neighbor コマンドで確認します。

```
AX620R1(config)#show ipv6 ospf neighbor
Neighbor ID     PID   Pri   State       Age   Uptime    Interface
10.0.0.1         1     1    FULL/DR      2    0:17:00   FastEthernet0/1.0
192.168.1.1      1     1    FULL/DR      6    0:16:57   FastEthernet1/0.0
```

■ AX620R2：

show ipv6 ospf neighbor コマンドで確認します。

```
AX620R2(config)#show ipv6 ospf neighbor
Neighbor ID     PID   Pri   State       Age   Uptime    Interface
172.16.1.1       1     1    FULL/BDR     1    0:22:28   FastEthernet0/1.0
10.0.0.1         1     1    FULL/DR      5    0:51:17   FastEthernet1/0.0
```

STEP11　OSPFv3 データベースの確認

さらに各ルーターの OSPFv3 データベースの内容を **show ipv6 ospf database** コマンドで確認します。

■ Cisco1812J：

show ipv6 ospf database コマンドで確認します。

```
Cisco1812J#show ipv6 ospf database

            OSPFv3 Router with ID (10.0.0.1) (Process ID 1)

                Router Link States (Area 0)

ADV Router      Age         Seq#            Fragment ID   Link count   Bits
10.0.0.1        322         0x8000002E      0             2            None
172.16.1.1      990         0x8000004D      0             2            None
192.168.1.1     995         0x8000002E      0             2            E

                Net Link States (Area 0)

ADV Router      Age         Seq#            Link ID       Rtr count
10.0.0.1        820         0x80000002      2             2
10.0.0.1        1582        0x80000003      3             2
172.16.1.1      990         0x80000003      3             2

                Link (Type-8) Link States (Area 0)

ADV Router      Age         Seq#            Link ID       Interface
10.0.0.1        322         0x80000003      20            Vl1
10.0.0.1        322         0x80000008      3             Fa1
192.168.1.1     1009        0x80000005      3             Fa1
10.0.0.1        322         0x80000007      2             Fa0
172.16.1.1      980         0x80000006      2             Fa0
```

09-1 マルチベンダー機器によるOSPFv3のデフォルトルートの伝搬

```
                   Intra Area Prefix Link States (Area 0)

ADV Router       Age         Seq#           Link ID    Ref-lstype   Ref-LSID
10.0.0.1         323         0x80000002     0          0x2001       0
10.0.0.1         820         0x80000002     2048       0x2002       2
10.0.0.1         1583        0x80000003     3072       0x2002       3
172.16.1.1       991         0x80000039     0          0x2001       0
172.16.1.1       977         0x80000006     3          0x2002       3


                   Type-5 AS External Link States

ADV Router       Age         Seq#           Prefix
192.168.1.1      1006        0x80000001     ::/0
```

■ AX620R1：

show ipv6 ospf database コマンドで確認します。

```
AX620R1(config)#show ipv6 ospf database

OSPFv3 router with process ID 1
Router ID 172.16.1.1

  Link LSAs for Interface FastEthernet0/0.0(ID 1)
  Adv. router        Age      Seq. No.      LSID         Priority    Prefixes
  172.16.1.1         1074     0x80000004    1            1           1

  Link LSAs for Interface FastEthernet0/1.0(ID 2)
  Adv. router        Age      Seq. No.      LSID         Priority    Prefixes
  10.0.0.1           460      0x80000007    2            1           1
  172.16.1.1         1115     0x80000006    2            1           1

  Link LSAs for Interface FastEthernet1/0.0(ID 3)
  Adv. router        Age      Seq. No.      LSID         Priority    Prefixes
  172.16.1.1         1112     0x80000004    3            1           1
  192.168.1.1        1137     0x80000003    2            1           1

  Router LSAs for Area 0.0.0.0(0)
  Adv. router        Age      Seq. No.      LSID         Link        Flags
  10.0.0.1           461      0x8000002e    0            2
  172.16.1.1         1127     0x8000004d    0            2
  192.168.1.1        1138     0x8000002e    0            2           E

  Network LSAs for Area 0.0.0.0(0)
  Adv. router        Age      Seq. No.      LSID         Routers
  10.0.0.1           960      0x80000002    2            2
  10.0.0.1           1722     0x80000003    3            2
  172.16.1.1         1128     0x80000003    3            2

  Intra-Area-Prefix LSAs for Area 0.0.0.0(0)
  Adv. router        Age      Seq. No.      LSID         Prefixes    Reftype    RefID
  10.0.0.1           462      0x80000002    0            1           Router     0
  10.0.0.1           960      0x80000002    2048         1           Network    2
  10.0.0.1           1722     0x80000003    3072         1           Network    3
  172.16.1.1         1128     0x80000039    0            1           Router     0
  172.16.1.1         1113     0x80000006    3            1           Network    3

  AS-External LSAs
  Adv. router        Age      Seq. No.      LSID         Prefix
  192.168.1.1        1144     0x80000001    0
```

Chapter 09 OSPFv3のデフォルトルートの伝搬

■ AX620R2:

`show ipv6 ospf database` コマンドで確認します。

```
AX620R2(config)#show ipv6 ospf database

OSPFv3 router with process ID 1
Router ID 192.168.1.1

  Link LSAs for Interface FastEthernet0/1.0(ID 2)
  Adv. router       Age      Seq. No.      LSID    Priority   Prefixes
  172.16.1.1        1214     0x80000004    3       1          1
  192.168.1.1       1238     0x80000003    2       1          1

  Link LSAs for Interface FastEthernet1/0.0(ID 3)
  Adv. router       Age      Seq. No.      LSID    Priority   Prefixes
  10.0.0.1          562      0x80000008    3       1          1
  192.168.1.1       1247     0x80000005    3       1          1

  Router LSAs for Area 0.0.0.0(0)
  Adv. router       Age      Seq. No.      LSID    Link       Flags
  10.0.0.1          562      0x8000002e    0       2
  172.16.1.1        1229     0x8000004d    0       2
  192.168.1.1       1238     0x8000002e    0       2          E

  Network LSAs for Area 0.0.0.0(0)
  Adv. router       Age      Seq. No.      LSID    Routers
  10.0.0.1          1061     0x80000002    2       2
  10.0.0.1          1823     0x80000003    3       2
  172.16.1.1        1230     0x80000003    3       2

  Intra-Area-Prefix LSAs for Area 0.0.0.0(0)
  Adv. router       Age      Seq. No.      LSID    Prefixes   Reftype   RefID
  10.0.0.1          563      0x80000002    0       1          Router    0
  10.0.0.1          1061     0x80000002    2048    1          Network   2
  10.0.0.1          1823     0x80000003    3072    1          Network   3
  172.16.1.1        1230     0x80000039    0       1          Router    0
  172.16.1.1        1216     0x80000006    3       1          Network   3

  AS-External LSAs
  Adv. router       Age      Seq. No.      LSID    Prefix
  192.168.1.1       1244     0x80000001    0       ::/0
```

STEP12 ルーティングテーブルの確認

`show ipv6 route` コマンドを用いて，ルーティングテーブル情報を表示し，デフォルトルートが表示されることを確認してください。

■ Cisco1812J:

`show ipv6 route` コマンドで確認します。

```
Cisco1812J#show ipv6 route
IPv6 Routing Table - 10 entries
Codes: C - Connected, L - Local, S - Static, R - RIP, B - BGP
       U - Per-user Static route, M - MIPv6
       I1 - ISIS L1, I2 - ISIS L2, IA - ISIS interarea, IS - ISIS summary
       O - OSPF intra, OI - OSPF inter, OE1 - OSPF ext 1, OE2 - OSPF ext 2
       ON1 - OSPF NSSA ext 1, ON2 - OSPF NSSA ext 2
       D - EIGRP, EX - EIGRP external
```

```
OE2     ::/0 [110/1]
          via FE80::260:B9FF:FE4F:223F, FastEthernet1
O       2001:DB8:0:1::/64 [110/2]
          via FE80::260:B9FF:FE4B:A4F7, FastEthernet0
C       2001:DB8:0:2::/64 [0/0]
          via ::, FastEthernet0
L       2001:DB8:0:2::2/128 [0/0]
          via ::, FastEthernet0
C       2001:DB8:0:3::/64 [0/0]
          via ::, Vlan1
L       2001:DB8:0:3::1/128 [0/0]
          via ::, Vlan1
O       2001:DB8:1:1::/64 [110/2]
          via FE80::260:B9FF:FE4F:223F, FastEthernet1
          via FE80::260:B9FF:FE4B:A4F7, FastEthernet0
C       2001:DB8:2::/64 [0/0]
          via ::, FastEthernet1
L       2001:DB8:2::1/128 [0/0]
          via ::, FastEthernet1
L       FF00::/8 [0/0]
          via ::, Null0
```

■ AX620R1：

`show ipv6 route` コマンドで確認します。

```
AX620R1(config)#show ipv6 route
IPv6 Routing Table - 16 entries, unlimited
Codes: C - Connected, L - Local, S - Static
       R - RIPng, O - OSPF, IA - OSPF inter area
       E1 - OSPF external type 1, E2 - OSPF external type 2, B - BGP
       s - Summary
Timers: Uptime/Age
O E2    ::/0 orphan [110/1]
          via fe80::260:b9ff:fe4f:22df, FastEthernet1/0.0, 0:08:06/0:00:00
C       2001:db8:0:1::/64 global [0/1]
          via ::, FastEthernet0/0.0, 1:12:45/0:00:00
O       2001:db8:0:1::/64 global [110/1]
          via ::, FastEthernet0/0.0, 0:08:07/0:00:00
L       2001:db8:0:1::/128 global [0/1]
          via ::, FastEthernet0/0.0, 1:12:46/0:00:00
L       2001:db8:0:1::1/128 global [0/1]
          via ::, FastEthernet0/0.0, 1:12:45/0:00:00
C       2001:db8:0:2::/64 global [0/1]
          via ::, FastEthernet0/1.0, 1:12:45/0:00:00
O       2001:db8:0:2::/64 global [110/1]
          via ::, FastEthernet0/1.0, 0:08:07/0:00:00
L       2001:db8:0:2::/128 global [0/1]
          via ::, FastEthernet0/1.0, 1:12:46/0:00:00
L       2001:db8:0:2::1/128 global [0/1]
          via ::, FastEthernet0/1.0, 1:12:46/0:00:00
O       2001:db8:0:3::/64 global [110/2]
          via fe80::226:bff:fef6:b5fa, FastEthernet0/1.0, 0:08:08/0:00:00
C       2001:db8:1:1::/64 global [0/1]
          via ::, FastEthernet1/0.0, 1:12:46/0:00:00
O       2001:db8:1:1::/64 global [110/1]
          via ::, FastEthernet1/0.0, 0:08:08/0:00:00
L       2001:db8:1:1::/128 global [0/1]
          via ::, FastEthernet1/0.0, 1:12:47/0:00:00
L       2001:db8:1:1::1/128 global [0/1]
```

Chapter 09 OSPFv3のデフォルトルートの伝搬

```
                via ::, FastEthernet1/0.0, 1:12:46/0:00:00
O      2001:db8:2::0/64 global [110/2]
                via fe80::226:bff:fef6:b5fa, FastEthernet0/1.0, 0:08:08/0:00:00
                via fe80::260:b9ff:fe4f:22df, FastEthernet1/0.0, 0:08:08/0:00:00
```

■ **AX620R2：**

show ipv6 route コマンドで確認します。

```
AX620R2(config)#show ipv6 route
IPv6 Routing Table - 17 entries, unlimited
Codes: C - Connected, L - Local, S - Static
       R - RIPng, O - OSPF, IA - OSPF inter area
       E1 - OSPF external type 1, E2 - OSPF external type 2, B - BGP
       s - Summary
Timers: Uptime/Age
S      ::/0 orphan [1/1]
                via ::, FastEthernet0/0.0, 0:22:19/0:00:00
O      ::/0 orphan [110/1]
                via ::, Null0.0, 0:21:43/0:00:00
O      2001:db8:0:1::/64 global [110/2]
                via fe80::260:b9ff:fe4b:a40f, FastEthernet0/1.0, 0:09:49/0:00:00
O      2001:db8:0:2::/64 global [110/2]
                via fe80::226:bff:fef6:b5fb, FastEthernet1/0.0, 0:09:49/0:00:00
                via fe80::260:b9ff:fe4b:a40f, FastEthernet0/1.0, 0:09:49/0:00:00
O      2001:db8:0:3::/64 global [110/2]
                via fe80::226:bff:fef6:b5fb, FastEthernet1/0.0, 0:09:49/0:00:00
C      2001:db8:1:1::/64 global [0/1]
                via ::, FastEthernet0/1.0, 1:49:09/0:00:00
O      2001:db8:1:1::/64 global [110/1]
                via ::, FastEthernet0/1.0, 0:09:49/0:00:00
L      2001:db8:1:1::/128 global [0/1]
                via ::, FastEthernet0/1.0, 1:49:10/0:00:00
L      2001:db8:1:1::2/128 global [0/1]
                via ::, FastEthernet0/1.0, 1:49:10/0:00:00
C      2001:db8:1:2::/64 global [0/1]
                via ::, FastEthernet0/0.0, 1:49:49/0:00:00
L      2001:db8:1:2::/128 global [0/1]
                via ::, FastEthernet0/0.0, 1:49:50/0:00:00
L      2001:db8:1:2::1/128 global [0/1]
                via ::, FastEthernet0/0.0, 1:49:49/0:00:00
C      2001:db8:2::/64 global [0/1]
                via ::, FastEthernet1/0.0, 1:49:28/0:00:00
O      2001:db8:2::/64 global [110/1]
                via ::, FastEthernet1/0.0, 0:09:50/0:00:00
L      2001:db8:2::/128 global [0/1]
                via ::, FastEthernet1/0.0, 1:49:29/0:00:00
L      2001:db8:2::2/128 global [0/1]
                via ::, FastEthernet1/0.0, 1:49:29/0:00:00
```

■実習終了時設定内容

■Cisco1821J：

```
version 12.4
service timestamps debug datetime msec
service timestamps log datetime msec
no service password-encryption
!
hostname Cisco1812J
!
boot-start-marker
boot-end-marker
!
no aaa new-model
!
dot11 syslog
!
ip cef
!
ipv6 unicast-routing
multilink bundle-name authenticated
!
archive
 log config
  hidekeys
!
interface FastEthernet0
 no ip address
 duplex auto
 speed auto
 ipv6 address 2001:DB8:0:2::2/64
 ipv6 enable
 ipv6 ospf 1 area 0
!
interface FastEthernet1
 no ip address
 duplex auto
 speed auto
 ipv6 address 2001:DB8:2::1/64
 ipv6 enable
 ipv6 ospf 1 area 0
!
interface BRI0
 no ip address
 encapsulation hdlc
 shutdown
!
interface FastEthernet2
!
interface FastEthernet3
!
interface FastEthernet4
!
interface FastEthernet5
!
interface FastEthernet6
!
interface FastEthernet7
!
```

Chapter 09 OSPFv3のデフォルトルートの伝搬

```
interface FastEthernet8
!
interface FastEthernet9
!
interface Vlan1
 no ip address
 ipv6 address 2001:DB8:0:3::1/64
 ipv6 enable
 ipv6 ospf 1 area 0
!
ip forward-protocol nd
!
no ip http server
no ip http secure-server
!
ipv6 router ospf 1
 router-id 10.0.0.1
 log-adjacency-changes
!
control-plane
!
line con 0
line aux 0
line vty 0 4
 login
!
end
```

■ **AX620R1：**

```
hostname AX620R1
timezone +09 00
!
ipv6 router ospf 1
  router-id 172.16.1.1
  area 0
  network FastEthernet0/0.0 area 0
  network FastEthernet0/1.0 area 0
  network FastEthernet1/0.0 area 0
!
device FastEthernet0/0
!
device FastEthernet0/1
!
device FastEthernet1/0
!
device BRI1/0
  isdn switch-type hsd128k
!
interface FastEthernet0/0.0
  no ip address
  ipv6 enable
  ipv6 address 2001:db8:0:1::1/64
  no shutdown
!
interface FastEthernet0/1.0
  no ip address
  ipv6 enable
  ipv6 address 2001:db8:0:2::1/64
  no shutdown
```

```
!
interface FastEthernet1/0.0
  no ip address
  ipv6 enable
  ipv6 address 2001:db8:1:1::1/64
  no shutdown
!
interface BRI1/0.0
  encapsulation ppp
  no auto-connect
  no ip address
  shutdown
!
interface Loopback0.0
  no ip address
!
interface Null0.0
  no ip address
```

■ **AX620R2：**

```
hostname AX620R2
timezone +09 00
!
ipv6 route default FastEthernet0/0.0
!
ipv6 router ospf 1
  router-id 192.168.1.1
  originate-default
  area 0
  network FastEthernet1/0.0 area 0
  network FastEthernet0/1.0 area 0
!
device FastEthernet0/0
!
device FastEthernet0/1
!
device FastEthernet1/0
!
device BRI1/0
  isdn switch-type hsd128k
!
interface FastEthernet0/0.0
  no ip address
  ipv6 enable
  ipv6 address 2001:db8:1:2::1/64
  no shutdown
!
interface FastEthernet0/1.0
  no ip address
  ipv6 enable
  ipv6 address 2001:db8:1:1::2/64
  no shutdown
!
interface FastEthernet1/0.0
  no ip address
  ipv6 enable
  ipv6 address 2001:db8:2::2/64
  no shutdown
!
```

```
interface BRI1/0.0
  encapsulation ppp
  no auto-connect
  no ip address
  shutdown
!
interface Loopback0.0
  no ip address
!
interface Null0.0
  no ip address
```

09-2 シスコ機器によるOSPFv3のデフォルトルートの伝搬

IPv6を用いたネットワーク上でOSPFv3を動作させ，デフォルトルートを伝搬させる実習です。本実習では3台ともCiscoルーターを用います。

■実習トポロジー

Cisco1841CのFastEternet 0/0をデフォルトルートとします。この情報をOSPFv3へ伝搬させます。

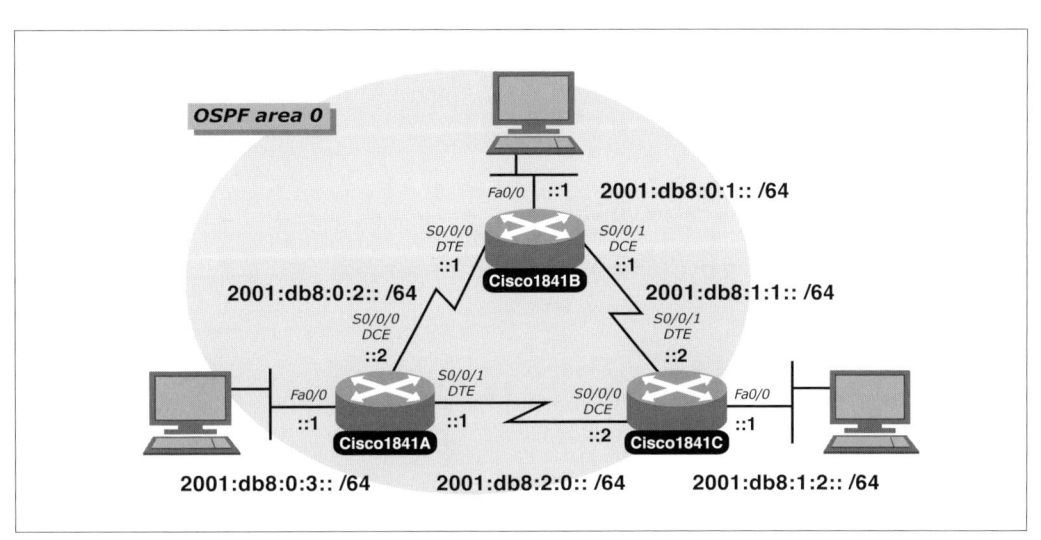

図9-2 ネットワークトポロジー

■インターフェース情報

表9-3 インターフェースアドレス割当一覧

ルーター	インターフェース	IPv6アドレス
Cisco1841A	FastEthernet 0/0 (Fa0/0)	2001:db8:0:3::1/64
Cisco1841A	Serial 0/0/0 (S0/0/0)	2001:db8:0:2::2/64
Cisco1841A	Serial 0/0/1 (S0/0/1)	2001:db8:2:0::1/64
Cisco1841B	FastEthernet 0/0 (Fa0/0)	2001:db8:0:1::1/64
Cisco1841B	Serial 0/0/0 (S0/0/0)	2001:db8:0:2::1/64
Cisco1841B	Serial 0/0/1 (S0/0/1)	2001:db8:1:1::1/64
Cisco1841C	FastEthernet 0/0 (Fa0/0)	2001:db8:1:2::1/64
Cisco1841C	Serial 0/0/0 (S0/0/0)	2001:db8:2:0::2/64
Cisco1841C	Serial 0/0/1 (S0/0/1)	2001:db8:1:1::2/64

Chapter 09 OSPFv3のデフォルトルートの伝搬

■ルーター ID 情報

表9-4 ルーター ID 情報一覧

ルーター	ルーター ID
Cisco1841A	10.0.0.1
Cisco1841B	172.16.1.1
Cisco1841C	192.168.1.1

■設定手順

STEP1 物理接続の実施

各ルーターおよび PC を "図9-2 ネットワークトポロジー" のとおりに接続します。シリアル回線における DCE/DTE は図内表記を参照してください。

STEP2 ホスト名の設定と IPv6 の有効化

Cisco ルーターの IPv6 トラフィック転送有効化とホスト名の設定を行います。

■Cisco1841A：

```
Router(config)#hostname Cisco1841A
Cisco1841A(config)#ipv6 unicast-routing
```

■Cisco1841B：

```
Router(config)#hostname Cisco1841B
Cisco1841B(config)#ipv6 unicast-routing
```

■Cisco1841C：

```
Router(config)#hostname Cisco1841C
Cisco1841C(config)#ipv6 unicast-routing
```

STEP3 インターフェースへの IPv6 アドレスの設定

すべてのルーターに IPv6 アドレスを設定し、インターフェースを有効化します。アドレスは "表9-3 インターフェースアドレス割当一覧" を参照してください。

■Cisco1841A：

```
Cisco1841A(config)#interface FastEthernet0/0
Cisco1841A(config-if)#ipv6 enable
Cisco1841A(config-if)#ipv6 address 2001:db8:0:3::1/64
Cisco1841A(config-if)#no shutdown
Cisco1841A(config)#interface Serial0/0/0
Cisco1841A(config-if)#ipv6 enable
Cisco1841A(config-if)#ipv6 address 2001:db8:0:2::2/64
Cisco1841A(config-if)#clock rate 64000
Cisco1841A(config-if)#no shutdown
Cisco1841A(config)#interface Serial0/0/1
Cisco1841A(config-if)#ipv6 enable
Cisco1841A(config-if)#ipv6 address 2001:db8:2::1/64
Cisco1841A(config-if)#no shutdown
```

■ Cisco1841B：

```
Cisco1841B(config)#interface FastEthernet0/0
Cisco1841B(config-if)#ipv6 enable
Cisco1841B(config-if)#ipv6 address 2001:db8:0:1::1/64
Cisco1841B(config-if)#no shutdown
Cisco1841B(config)#interface Serial0/0/0
Cisco1841B(config-if)#ipv6 enable
Cisco1841B(config-if)#ipv6 address 2001:db8:0:2::1/64
Cisco1841B(config-if)#no shutdown
Cisco1841B(config)#interface Serial0/0/1
Cisco1841B(config-if)#ipv6 enable
Cisco1841B(config-if)#ipv6 address 2001:db8:1:1::1/64
Cisco1841B(config-if)#clock rate 64000
Cisco1841B(config-if)#no shutdown
```

■ Cisco1841C：

```
Cisco1841C(config)#interface FastEthernet0/0
Cisco1841C(config-if)#ipv6 enable
Cisco1841C(config-if)#ipv6 address 2001:db8:1:2::1/64
Cisco1841C(config-if)#no shutdown
Cisco1841C(config)#interface Serial0/0/0
Cisco1841C(config-if)#ipv6 enable
Cisco1841C(config-if)#ipv6 address 2001:db8:2::2/64
Cisco1841C(config-if)#clock rate 64000
Cisco1841C(config-if)#no shutdown
Cisco1841C(config)#interface Serial0/0/1
Cisco1841C(config-if)#ipv6 enable
Cisco1841C(config-if)#ipv6 address 2001:db8:1:1::2/64
Cisco1841C(config-if)#no shutdown
```

STEP4 インターフェースへのアドレス確認

インターフェースのステータスと動作していることを確認します。

`show ipv6 interface brief` コマンドを用いてインターフェース情報を表示させます。下記に出力を示します。インターフェース名の後の表示が [up/up] なら動作しています。

■ Cisco1841A：

`show ipv6 interface brief` コマンドで確認します。

```
Cisco1841A#show ipv6 interface brief
FastEthernet0/0            [up/up]
    FE80::21D:70FF:FEE6:33D0
    2001:DB8:0:3::1
FastEthernet0/1            [administratively down/down]
Serial0/0/0                [up/up]
    FE80::21D:70FF:FEE6:33D0
    2001:DB8:0:2::2
Serial0/0/1                [up/up]
    FE80::21D:70FF:FEE6:33D0
    2001:DB8:2::1
```

■ Cisco1841B：

`show ipv6 interface brief` コマンドで確認します。

```
Cisco1841B#show ipv6 interface brief
FastEthernet0/0            [up/up]
```

```
    FE80::211:21FF:FE67:B480
    2001:DB8:0:1::1
Serial0/0/0                    [up/up]
    FE80::211:21FF:FE67:B480
    2001:DB8:0:2::1
FastEthernet0/1                [administratively down/down]
Serial0/0/1                    [up/up]
    FE80::211:21FF:FE67:B480
    2001:DB8:1:1::1
```

■ **Cisco1841C：**

`show ipv6 interface brief` コマンドで確認します。

```
Cisco1841C#show ipv6 interface brief
FastEthernet0/0                [up/up]
    FE80::21D:70FF:FEE6:3268
    2001:DB8:1:2::1
FastEthernet0/1                [administratively down/down]
Serial0/0/0                    [up/up]
    FE80::21D:70FF:FEE6:3268
    2001:DB8:2::2
Serial0/0/1                    [up/up]
    FE80::21D:70FF:FEE6:3268
    2001:DB8:1:1::2
```

STEP5　OSPFv3 プロセスの設定

　OSPFv3 をプロセス番号 1 で起動し，各ルーターにルーター ID を設定します。ルーター ID は " 表 9-4　ルーター ID 情報一覧 " を参照してください。

■ **Cisco1841A：**

```
Cisco1841A(config)#ipv6 router ospf 1
Cisco1841A(config-router)#router-id 10.0.0.1
```

■ **Cisco1841B：**

```
Cisco1841B(config)#ipv6 router ospf 1
Cisco1841B(config-router)#router-id 172.16.1.1
```

■ **Cisco1841C：**

```
Cisco1841C(config)#ipv6 router ospf 1
Cisco1841C(config-router)#router-id 192.168.1.1
```

STEP6　インターフェースへの OSPFv3 プロセス及びエリアの割当

　各インターフェースに OSPFv3 プロセス 1，エリア 0 を割り当てます。

■ **Cisco1841A：**

```
Cisco1841A(config)#interface FastEthernet0/0
Cisco1841A(config-if)#ipv6 ospf 1 area 0
Cisco1841A(config)#interface Serial0/0/0
Cisco1841A(config-if)#ipv6 ospf 1 area 0
Cisco1841A(config)#interface Serial0/0/1
Cisco1841A(config-if)#ipv6 ospf 1 area 0
```

09-2 シスコ機器によるOSPFv3のデフォルトルートの伝搬

■ Cisco1841B：

```
Cisco1841B(config)#interface FastEthernet0/0
Cisco1841B(config-if)#ipv6 ospf 1 area 0
Cisco1841B(config)#iinterface Serial0/0/0
Cisco1841B(config-if)#ipv6 ospf 1 area 0
Cisco1841B(config)#iinterface Serial0/0/1
Cisco1841B(config-if)#ipv6 ospf 1 area 0
```

■ Cisco1841C：

```
Cisco1841C(config)#interface Serial0/0/0
Cisco1841C(config-if)#ipv6 ospf 1 area 0
Cisco1841C(config)#interface Serial0/0/1
Cisco1841C(config-if)#ipv6 ospf 1 area 0
```

STEP7 デフォルトルートのインターフェースへの設定

Cisco1841C ルーター上でデフォルトルートを FastEthernet 0/0 に設定します。

■ Cisco1841C：

```
Cisco1841C(config)#ipv6 route ::/0 FastEthernet0/0
```

STEP8 デフォルトルートの OSPFv3 への伝搬

Cisco1841C ルーター上でデフォルトルートをアドバタイズさせます。

■ Cisco1841C：

```
Cisco1841C(config)#ipv6 router ospf 1
Cisco1841C(config-router)#default-information originate
```

STEP9 OSPFv3 設定確認

各ルーターで動作しているルーティングプロトコルが設定した内容 (OSPFv3 process1 Area0) であることを **show ipv6 protocols** コマンドで確認します。

■ Cisco1841A：

show ipv6 protocols コマンドで確認します。

```
Cisco1841A#show ipv6 protocols
IPv6 Routing Protocol is "connected"
IPv6 Routing Protocol is "static"
IPv6 Routing Protocol is "ospf 1"
  Interfaces (Area 0):
    Serial0/0/1
    Serial0/0/0
    FastEthernet0/0
  Redistribution:
    None
```

■ Cisco1841B：

show ipv6 protocols コマンドで確認します。

```
Cisco1841B#show ipv6 protocols
```

Chapter 09 OSPFv3のデフォルトルートの伝搬

```
  IPv6 Routing Protocol is "connected"
  IPv6 Routing Protocol is "static"
  IPv6 Routing Protocol is "ospf 1"
    Interfaces (Area 0):
      Serial0/0/1
      Serial0/0/0
      FastEthernet0/0
    Redistribution:
      None
```

■ Cisco1841C：

show ipv6 protocols コマンドで確認します。

```
Cisco1841C#show ipv6 protocols
IPv6 Routing Protocol is "connected"
IPv6 Routing Protocol is "static"
IPv6 Routing Protocol is "ospf 1"
  Interfaces (Area 0):
    Serial0/0/1
    Serial0/0/0
  Redistribution:
    None
```

STEP10 ルーター相互の隣接情報確認

各ルーターの隣接情報などを **show ipv6 ospf neighbor** コマンドで確認します。特に State と Interface ID を確認してください。

■ Cisco1841A：

show ipv6 ospf neighbor コマンドで確認します。

```
Cisco1841A#show ipv6 ospf neighbor

Neighbor ID     Pri   State          Dead Time   Interface ID    Interface
192.168.1.1       1   FULL/  -       00:00:37    6               Serial0/0/1
172.16.1.1        1   FULL/  -       00:00:32    6               Serial0/0/0
```

■ Cisco1841B：

show ipv6 ospf neighbor コマンドで確認します。

```
Cisco1841B#show ipv6 ospf neighbor

Neighbor ID     Pri   State          Dead Time   Interface ID    Interface
192.168.1.1       1   FULL/  -       00:00:33    7               Serial0/1
10.0.0.1          1   FULL/  -       00:00:38    6               Serial0/0
```

■ Cisco1841C：

show ipv6 ospf neighbor コマンドで確認します。

```
Cisco1841C#show ipv6 ospf neighbor

Neighbor ID     Pri   State          Dead Time   Interface ID    Interface
172.16.1.1        1   FULL/  -       00:00:35    7               Serial0/0/1
10.0.0.1          1   FULL/  -       00:00:38    7               Serial0/0/0
```

STEP11　OSPFv3 データーベースの確認

さらに各ルーターの OSPFv3 データベースの内容を **show ipv6 ospf database** コマンドで確認します。

■ Cisco1841A：

show ipv6 ospf database コマンドで確認します。

```
Cisco1841A#show ipv6 ospf database

            OSPFv3 Router with ID (10.0.0.1) (Process ID 1)

              Router Link States (Area 0)

ADV Router      Age         Seq#         Fragment ID  Link count  Bits
10.0.0.1        1237        0x80000007   0            2           None
172.16.1.1      1198        0x80000002   0            2           None
192.168.1.1     1235        0x80000006   0            2           E

              Link (Type-8) Link States (Area 0)

ADV Router      Age         Seq#         Link ID      Interface
10.0.0.1        1389        0x80000001   7            Se0/0/1
192.168.1.1     1388        0x80000001   6            Se0/0/1
10.0.0.1        1248        0x80000002   6            Se0/0/0
172.16.1.1      1238        0x80000001   6            Se0/0/0
10.0.0.1        1389        0x80000001   4            Fa0/0

              Intra Area Prefix Link States (Area 0)

ADV Router      Age         Seq#         Link ID      Ref-lstype   Ref-LSID
10.0.0.1        1247        0x80000004   0            0x2001       0
172.16.1.1      1238        0x80000001   0            0x2001       0
192.168.1.1     1246        0x80000004   0            0x2001       0

              Type-5 AS External Link States

ADV Router      Age         Seq#         Prefix
192.168.1.1     1394        0x80000001   ::/0
```

■ Cisco1841B：

show ipv6 ospf database コマンドで確認します。

```
Cisco1841B#show ipv6 ospf database

            OSPFv3 Router with ID (172.16.1.1) (Process ID 1)

              Router Link States (Area 0)

ADV Router      Age         Seq#         Fragment ID  Link count  Bits
10.0.0.1        1300        0x80000007   0            2           None
172.16.1.1      1259        0x80000002   0            2           None
192.168.1.1     1297        0x80000006   0            2           E

              Link (Type-8) Link States (Area 0)

ADV Router      Age         Seq#         Link ID      Interface
172.16.1.1      1299        0x80000001   7            Se0/0/1
```

Chapter 09 OSPFv3のデフォルトルートの伝搬

```
192.168.1.1      1297       0x80000002      7         Se0/0/1
10.0.0.1         1310       0x80000002      6         Se0/0/0
172.16.1.1       1299       0x80000001      6         Se0/0/0
172.16.1.1       1299       0x80000001      4         Fa0/0

            Intra Area Prefix Link States (Area 0)

ADV Router       Age        Seq#            Link ID   Ref-lstype      Ref-LSID
10.0.0.1         1310       0x80000004      0         0x2001          0
172.16.1.1       1299       0x80000001      0         0x2001          0
192.168.1.1      1300       0x80000004      0         0x2001          0

            Type-5 AS External Link States

ADV Router       Age        Seq#            Prefix
192.168.1.1      1450       0x80000001      ::/0
```

■ **Cisco1841C：**

show ipv6 ospf database コマンドで確認します。

```
Cisco1841C#show ipv6 ospf database

        OSPFv3 Router with ID (192.168.1.1) (Process ID 1)

            Router Link States (Area 0)

ADV Router       Age        Seq#            Fragment ID   Link count    Bits
10.0.0.1         1339       0x80000007      0             2             None
172.16.1.1       1299       0x80000002      0             2             None
192.168.1.1      1335       0x80000006      0             2             E

            Link (Type-8) Link States (Area 0)

ADV Router       Age        Seq#            Link ID   Interface
172.16.1.1       1339       0x80000001      7         Se0/0/1
192.168.1.1      1335       0x80000002      7         Se0/0/1
10.0.0.1         1491       0x80000001      7         Se0/0/0
192.168.1.1      1487       0x80000001      6         Se0/0/0

            Intra Area Prefix Link States (Area 0)

ADV Router       Age        Seq#            Link ID   Ref-lstype      Ref-LSID
10.0.0.1         1350       0x80000004      0         0x2001          0
172.16.1.1       1339       0x80000001      0         0x2001          0
192.168.1.1      1338       0x80000004      0         0x2001          0

            Type-5 AS External Link States

ADV Router       Age        Seq#            Prefix
192.168.1.1      1488       0x80000001      ::/0
```

STEP12 ルーティングテーブルの確認

show ipv6 route コマンドを用いて，ルーティングテーブル情報を表示し，デフォルトルートが表示されることを確認してください。

09-2 シスコ機器によるOSPFv3のデフォルトルートの伝搬

■ Cisco1841A：

show ipv6 route コマンドで確認します。

```
Cisco1841A#show ipv6 route
IPv6 Routing Table - Default - 10 entries
Codes: C - Connected, L - Local, S - Static, U - Per-user Static route
       B - BGP, M - MIPv6, R - RIP, I1 - ISIS L1
       I2 - ISIS L2, IA - ISIS interarea, IS - ISIS summary, D - EIGRP
       EX - EIGRP external
       O - OSPF Intra, OI - OSPF Inter, OE1 - OSPF ext 1, OE2 - OSPF ext 2
       ON1 - OSPF NSSA ext 1, ON2 - OSPF NSSA ext 2
OE2 ::/0 [110/1], tag 1
     via FE80::21D:70FF:FEE6:33D0, Serial0/0/1
O   2001:DB8:0:1::/64 [110/782]
     via FE80::211:21FF:FE67:B480, Serial0/0/0
C   2001:DB8:0:2::/64 [0/0]
     via Serial0/0/0, directly connected
L   2001:DB8:0:2::2/128 [0/0]
     via Serial0/0/0, receive
C   2001:DB8:0:3::/64 [0/0]
     via FastEthernet0/0, directly connected
L   2001:DB8:0:3::1/128 [0/0]
     via FastEthernet0/0, receive
O   2001:DB8:1:1::/64 [110/1562]
     via FE80::21D:70FF:FEE6:33D0, Serial0/0/1
     via FE80::211:21FF:FE67:B480, Serial0/0/0
C   2001:DB8:2::/64 [0/0]
     via Serial0/0/1, directly connected
L   2001:DB8:2::1/128 [0/0]
     via Serial0/0/1, receive
L   FF00::/8 [0/0]
     via Null0, receive
```

■ Cisco1841B：

show ipv6 route コマンドで確認します。

```
Cisco1841B#show ipv6 route
IPv6 Routing Table - 11 entries
Codes: C - Connected, L - Local, S - Static, R - RIP, B - BGP
       U - Per-user Static route
       I1 - ISIS L1, I2 - ISIS L2, IA - ISIS interarea, IS - ISIS summary
       O - OSPF intra, OI - OSPF inter, OE1 - OSPF ext 1, OE2 - OSPF ext 2
       ON1 - OSPF NSSA ext 1, ON2 - OSPF NSSA ext 2
OE2 ::/0 [110/1], tag 1
     via FE80::21D:70FF:FEE6:33D0, Serial0/1
C   2001:DB8:0:1::/64 [0/0]
     via ::, FastEthernet0/0
L   2001:DB8:0:1::1/128 [0/0]
     via ::, FastEthernet0/0
C   2001:DB8:0:2::/64 [0/0]
     via ::, Serial0/0/0
L   2001:DB8:0:2::1/128 [0/0]
     via ::, Serial0/0/0
O   2001:DB8:0:3::/64 [110/782]
     via FE80::21D:70FF:FEE6:3268, Serial0/0/0
C   2001:DB8:1:1::/64 [0/0]
     via ::, Serial0/0/1
L   2001:DB8:1:1::1/128 [0/0]
     via ::, Serial0/0/1
```

```
O   2001:DB8:2::/64 [110/1562]
      via FE80::21D:70FF:FEE6:33D0, Serial0/0/1
      via FE80::21D:70FF:FEE6:3268, Serial0/0/0
L   FE80::/10 [0/0]
      via ::, Null0
L   FF00::/8 [0/0]
      via ::, Null0
```

■ **Cisco1841C:**

show ipv6 route コマンドで確認します。

```
Cisco1841C#show ipv6 route
IPv6 Routing Table - Default - 11 entries
Codes: C - Connected, L - Local, S - Static, U - Per-user Static route
       B - BGP, M - MIPv6, R - RIP, I1 - ISIS L1
       I2 - ISIS L2, IA - ISIS interarea, IS - ISIS summary, D - EIGRP
       EX - EIGRP external
       O - OSPF Intra, OI - OSPF Inter, OE1 - OSPF ext 1, OE2 - OSPF ext 2
       ON1 - OSPF NSSA ext 1, ON2 - OSPF NSSA ext 2
S   ::/0 [1/0]
      via FastEthernet0/0, directly connected
O   2001:DB8:0:1::/64 [110/782]
      via FE80::211:21FF:FE67:B480, Serial0/0/1
O   2001:DB8:0:2::/64 [110/1562]
      via FE80::21D:70FF:FEE6:3268, Serial0/0/0
      via FE80::211:21FF:FE67:B480, Serial0/0/1
O   2001:DB8:0:3::/64 [110/782]
      via FE80::21D:70FF:FEE6:3268, Serial0/0/0
C   2001:DB8:1:1::/64 [0/0]
      via Serial0/0/1, directly connected
L   2001:DB8:1:1::2/128 [0/0]
      via Serial0/0/1, receive
C   2001:DB8:1:2::/64 [0/0]
      via FastEthernet0/0, directly connected
L   2001:DB8:1:2::1/128 [0/0]
      via FastEthernet0/0, receive
C   2001:DB8:2::/64 [0/0]
      via Serial0/0/0, directly connected
L   2001:DB8:2::2/128 [0/0]
      via Serial0/0/0, receive
L   FF00::/8 [0/0]
      via Null0, receive
```

■実習終了時設定内容

■ Cisco1841A:

```
!
version 12.4
service timestamps debug datetime msec
service timestamps log datetime msec
no service password-encryption
!
hostname Cisco1841A
!
boot-start-marker
boot-end-marker
!
logging message-counter syslog
!
no aaa new-model
dot11 syslog
ip source-route
!
ip cef
ipv6 unicast-routing
ipv6 cef
!
multilink bundle-name authenticated
!
archive
 log config
  hidekeys
!
interface FastEthernet0/0
 no ip address
 duplex auto
 speed auto
 ipv6 address 2001:DB8:0:3::1/64
 ipv6 ospf 1 area 0
 no shutdown
!
interface FastEthernet0/1
 no ip address
 shutdown
 duplex auto
 speed auto
!
interface Serial0/0/0
 no ip address
 ipv6 address 2001:DB8:0:2::2/64
 ipv6 ospf 1 area 0
 clock rate 64000
 no shutdown
!
interface Serial0/0/1
 no ip address
 ipv6 address 2001:DB8:2::1/64
 ipv6 enable
 ipv6 ospf 1 area 0
 no shutdown
!
ip forward-protocol nd
```

```
!
!
ip http server
no ip http secure-server
!
ipv6 router ospf 1
 router-id 10.0.0.1
 log-adjacency-changes
!
control-plane
!
line con 0
line aux 0
line vty 0 4
 login
!
scheduler allocate 20000 1000
!
end
```

■ **Cisco1841B：**

```
!
version 12.4
service timestamps debug datetime msec
service timestamps log datetime msec
no service password-encryption
!
hostname Cisco1841B
!
boot-start-marker
boot-end-marker
!
logging message-counter syslog
!
no aaa new-model
dot11 syslog
ip source-route
!
ip cef
ipv6 unicast-routing
ipv6 cef
!
multilink bundle-name authenticated
!
archive
 log config
  hidekeys
!
interface FastEthernet0/0
 no ip address
 duplex auto
 speed auto
 ipv6 address 2001:DB8:0:1::1/64
 ipv6 ospf 1 area 0
 no shutdown
!
interface Serial0/0/0
 no ip address
 ipv6 address 2001:DB8:0:2::1/64
```

```
 ipv6 ospf 1 area 0
 no fair-queue
 no shutdown
!
interface FastEthernet0/0/1
 no ip address
 shutdown
 duplex auto
 speed auto
!
interface Serial0/0/1
 no ip address
 ipv6 address 2001:DB8:1:1::1/64
 ipv6 ospf 1 area 0
 clock rate 64000
 no shutdown
!
ip forward-protocol nd
!
ip http server
no ip http secure-server
!
ipv6 router ospf 1
 router-id 172.16.1.1
 log-adjacency-changes
!
control-plane
!
line con 0
line aux 0
line vty 0 4
 login
!
end
```

■ Cisco1841C：

```
version 12.4
service timestamps debug datetime msec
service timestamps log datetime msec
no service password-encryption
!
hostname Cisco1841C
!
boot-start-marker
boot-end-marker
!
logging message-counter syslog
!
no aaa new-model
memory-size iomem 10
!
dot11 syslog
ip source-route
!
ip cef
!
ipv6 unicast-routing
ipv6 cef
!
```

```
multilink bundle-name authenticated
!
voice-card 0
!
archive
 log config
  hidekeys
!
interface FastEthernet0/0
 no ip address
 duplex auto
 speed auto
 ipv6 address 2001:DB8:1:2::1/64
!
interface FastEthernet0/1
 no ip address
 shutdown
 duplex auto
 speed auto
!
interface Serial0/0/0
 no ip address
 ipv6 address 2001:DB8:2::2/64
 ipv6 enable
 ipv6 ospf 1 area 0
 clock rate 64000
!
interface Serial0/0/1
 no ip address
 ipv6 address 2001:DB8:1:1::2/64
 ipv6 ospf 1 area 0
!
ip forward-protocol nd
ip http server
no ip http secure-server
!
ipv6 route ::/0 FastEthernet0/0
ipv6 router ospf 1
 router-id 192.168.1.1
 log-adjacency-changes
 default-information originate
!
control-plane
!
mgcp fax t38 ecm
!
line con 0
line aux 0
line vty 0 4
 login
!
scheduler allocate 20000 1000
end
```

Chapter 10

IPv6 活用

10-1 IPv6 over IPv4
　　 トンネリングの設定

10-2 IPv6実習サーバーの
　　 構築

Chapter 10 IPv6活用

10-1 IPv6 over IPv4トンネリングの設定

　この章では3台のルーターを使ってIPv6 over IPv4 トンネリングを学習します。この技術によってISPなどがIPv6 サービスを提供していなくても拠点間相互にはIPv6 の運用が可能になります。
　トポロジーは以下のとおりです。

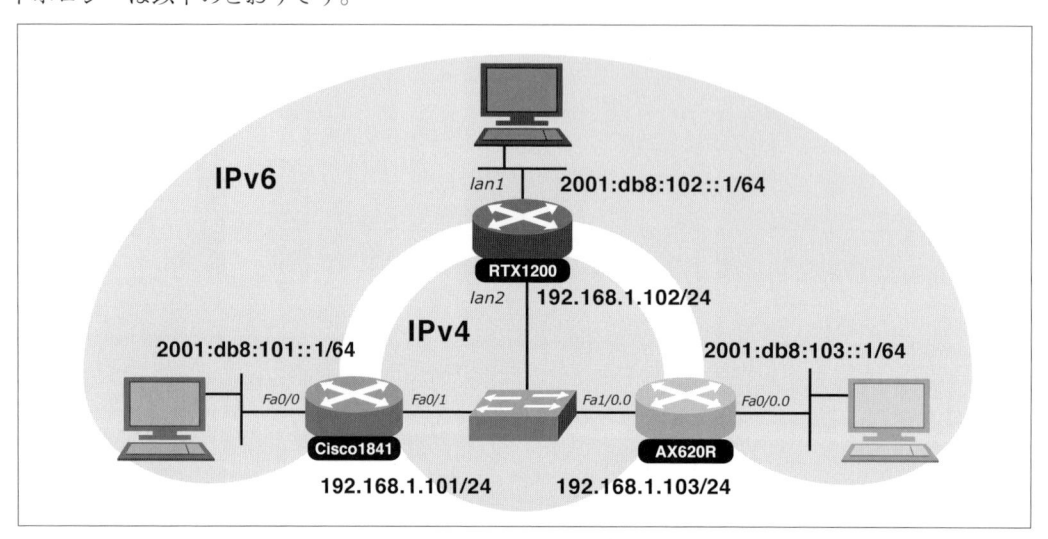

図10-1　ネットワークトポロジー

アドレスの割当は以下のとおりです。

表10-1　インターフェースアドレス割当一覧

ルーター	インターフェース	IPv6アドレス
Cisco1841	FastEthernet 0/0 (Fa0/0)	2001:db8:101::1/64
Cisco1841	FastEthernet 0/1 (Fa0/1)	192.168.1.101/24
RTX1200	lan1	2001:db8:102::1/64
RTX1200	lan2	192.168.1.102/24
AX620R	FastEthernet 0/0.0 (Fa0/0.0)	2001:db8:103::1/64
AX620R	FastEthernet 1/0.0 (Fa1/0.0)	192.168.1.103/24

STEP1　ルーターの初期化

各ルーターの設定に入る前に，事前準備を行います。まずはルーターの初期化です。

■ Cisco1841:

```
Router#erase startup-config
Router#reload
```

■ RTX1200:

次のコマンドで再起動すると，工場出荷時の設定で起動します。

```
#cold start
```

RTX1200は工場出荷時の設定ではLAN1にIPv6アドレスが割り当てられDHCPサーバーが起動しています。以下のコマンドでDHCPサーバーを停止し，IPv6アドレスの設定を消去します。

```
#no dhcp service server
#no ip lan1 address 192.168.100.1/24
```

■ AX620R:

```
Router(config)#erase startup-config
Router(config)#exit
Router#restart
```

次にホスト名の設定をします。

■ Cisco1841:

```
Router(config)#hostname Cisco1841
Cisco1841(config)#
```

■ RTX1200:

ホスト名の設定はできませんが，コマンドプロンプトの変更は可能です。

```
#console prompt RTX1200
RTX1200#
```

■ AX620R:

```
Router(config)#hostname AX620R
AX620R(config)#
```

■ Cisco1841:

シスコルーターは IPv6 の有効化を行う必要があります。

```
Cisco1841(config)#ipv6 unicast-routing
```

STEP2 物理接続の実施

各ルーターおよび PC を "図 10-1　ネットワークトポロジー" のとおりに接続します。

STEP3 インターフェースへの IPv6 および IPv4 の設定

各ルーターのそれぞれのインターフェースに IPv6 および IPv4 アドレスを設定します。

■ Cisco1841:

```
Cisco1841(config)#interface fastEthernet 0/0
Cisco1841(config-if)#ipv6 enable
Cisco1841(config-if)#ipv6 address 2001:db8:101::1/64
Cisco1841(config-if)#no shutdown
Cisco1841(config-if)#exit

Cisco1841(config)#interface fastEthernet 0/1
Cisco1841(config-if)#ip address 192.168.1.101 255.255.255.0
Ciaco1841(config-if)#no shutdown
```

■ RTX1200:

```
RTX1200#ipv6 lan1 address 2001:db8:102::1/64
RTX1200#ip lan2 address 192.168.1.102/24
```

Chapter 10 IPv6活用

■ **AX620R：**

```
AX620R(config)#interface FastEthernet0/0.0
AX620R(config-FastEthernet0/0.0)#ipv6 enable
AX620R(config-FastEthernet0/0.0)#ipv6 address 2001:db8:103::1/64
AX620R(config-FastEthernet0/0.0)#no shutdown

AX620R(config)#interface FastEthernet1/0.0
AX620R(config-FastEthernet1/0.0)#ip address 192.168.1.103/24
AX620R(config-FastEthernet1/0.0)#no shutdown
AX620R(config-FastEthernet1/0.0)#exit
```

STEP4　ルーター相互の通信確認

ルーター同士がIPv4で相互に通信できることを確認します。

■ **Cisco1841：**

RTX1200およびAX620Rに対して **ping** コマンドを実行します。

```
Cisco1841#ping 192.168.1.102
Cisco1841#ping 192.168.1.103
```

■ **RTX1200：**

Cisco1841およびAX620Rに対して **ping** コマンドを実行します。**Ping** コマンドは実行し続けますので中止するときはCtl+cで止めます。

```
RTX1200#ping 192.168.1.101
RTX1200#ping 192.168.1.103
```

■ **AX620R：**

Cisco1841およびRTX1200に対して **ping** コマンドを実行します。

```
AX620R(config)#ping 192.168.1.101
AX620R(config)#ping 192.168.1.102
```

STEP5　トンネルの設定

IPv6 over IPv4 トンネルを設定します。

■ **Cisco1841：**

トンネルインターフェースを作成し，トンネルを張る相手の指定をします。

まず，RTX1200に対するトンネルインターフェースを作成します。

```
Cisco1841(config)#interface tunnel0
Cisco1841(config-if)#ipv6 enable
Cisco1841(config-if)#tunnel mode ipv6ip
Cisco1841(config-if)#tunnel source fastethrnet0/1
Cisco1841(config-if)#tunnel destination 192.168.1.102
```

次にAX620Rに対するトンネルインターフェースを作成します。

```
Cisco1841(config)#interface tunnel1
Cisco1841(config-if)#ipv6 enable
Cisco1841(config-if)#tunnel mode ipv6ip
Cisco1841(config-if)#tunnel source fastethrnet0/1
Cisco1841(config-if)#tunnel destination 192.168.1.103
```

10-1 IPv6 over IPv4トンネリングの設定

■RTX1200:

Cisco1841に対するトンネルを設定します。

```
RTX1200#tunnel select 1
RTX1200tunnel 1#tunnel encapsulation ipip
RTX1200tunnel 1#tunnel endpoint address 192.168.1.101
RTX1200tunnel 1#tunnel enable 1
RTX1200tunnel 1#tunnel select none
```

次にAX620Rに対するトンネルを設定します。

```
RTX1200#tunnel select 2
RTX1200tunnel 2#tunnel encapsulation ipip
RTX1200tunnel 2#tunnel endpoint address 192.168.1.103
RTX1200tunnel 2#tunnel enable 2
RTX1200tunnel 2#tunnel select none
```

■AX620R:

Cisco1841と同様にトンネルインターフェースを作成し、トンネルを張る相手の指定をします。

まずCisco1841に対して、以下のとおりに入力します。

```
AX620R(config)#interface tunnel0.0
AX620R(config-Tunnel0.0)#tunnel mode 6-over-4
AX620R(config-Tunnel0.0)#tunnel destination 192.168.1.101
AX620R(config-Tunnel0.0)#tunnel source FastEthernet1/0.0
AX620R(config-Tunnel0.0)#ipv6 address autoconfig
AX620R(config-Tunnel0.0)#ipv6 enable
AX620R(config-Tunnel0.0)#no shutdown
```

次にRTX1200に対して、以下のとおりに入力します。

```
AX620R(config)#interface tunnel1.0
AX620R(config-Tunnel1.0)#tunnel mode 6-over-4
AX620R(config-Tunnel1.0)#tunnel destination 192.168.1.102
AX620R(config-Tunnel1.0)#tunnel source FastEthernet1/0.0
AX620R(config-Tunnel1.0)#ipv6 address autoconfig
AX620R(config-Tunnel1.0)#ipv6 enable
AX620R(config-Tunnel1.0)#no shutdown
```

STEP6 RIPngの設定

RIPngを用いてIPv6をルーティングします。

■Cisco1841:

```
Cisco1841(config)#ipv6 router rip RIPng
Cisco1841(config-rtr)#interface fastethrnet0/0
Cisco1841(config-if)#ipv6 rip RIPng enable
Cisco1841(config)#interface tunnel0
Cisco1841(config-if)#ipv6 rip RIPng enable
Cisco1841(config)#interface tunnel1
Cisco1841(config-if)#ipv6 rip RIPng enable
```

■RTX1200:

```
RTX1200#ipv6 rip use on
```

■ AX620R:

```
AX620R(config)#ipv6 router rip
AX620R(config)#interface FastEthernet0/0.0
AX620R(config-FastEthernet0/0.0)#ipv6 rip enable
AX620R(config)#interface Tunnel0.0
AX620R(config-Tunnel0.0)#ipv6 rip enable
AX620R(config)#interface Tunnel1.0
AX620R(config-Tunnel1.0)#ipv6 rip enable
```

STEP7 通信の確認

ルーター相互に他ルーターの IPv6 アドレスに **ping** コマンドを実行して通信を確認します。

■ Cisco1841:

```
Cisco1841#ping 2001:db8:102::1
Cisco1841#ping 2001:db8:103::1
```

■ RTX1200:

```
RTX1200#ping6 2001:db8:101::1
RTX1200#ping6 2001:db8:103::1
```

■ AX620R:

```
AX620R(config)#ping 2001:db8:101::1
AX620R(config)#ping 2001:db8:102::1
```

STEP8 ルーティングテーブルの確認

各ルーターの IPv6 のルーティングテーブルを確認し，他ルーターからの経路が RIPng により受け取れているかを確認します。

■ Cisco1841:

show ipv6 route コマンドで確認します。

```
Cisco1841#show ipv6 route
IPv6 Routing Table - 6 entries
Codes: C - Connected, L - Local, S - Static, R - RIP, B - BGP
       U - Per-user Static route
       I1 - ISIS L1, I2 - ISIS L2, IA - ISIS interarea, IS - ISIS summary
       O - OSPF intra, OI - OSPF inter, OE1 - OSPF ext 1, OE2 - OSPF ext 2
       ON1 - OSPF NSSA ext 1, ON2 - OSPF NSSA ext 2
C   2001:DB8:101::/64 [0/0]
     via ::, FastEthernet0/0
L   2001:DB8:101::1/128 [0/0]
     via ::, FastEthernet0/0
R   2001:DB8:102::/64 [120/2]
     via FE80::2A0:DEFF:FE65:AFD1, Tunnel0
R   2001:DB8:103::/64 [120/2]
     via FE80::2F3:F3FF:FEB6:E588, Tunnel1
L   FE80::/10 [0/0]
     via ::, Null0
L   FF00::/8 [0/0]
     via ::, Null0
Cisco1841#
```

■ RTX1200:

show ipv6 route コマンドで確認します。

```
RTX1200#show ipv6 route
Destination            Gateway                    Interface   Type
2001:db8:101::/64      fe80::c0a8:165             TUNNEL[1]   RIPng
2001:db8:102::/64      -                          LAN1        implicit
2001:db8:103::/64      fe80::2f4:f7ff:feb6:e588   TUNNEL[2]   RIPng
RTX1200#
```

■ AX620R:

show ipv6 route コマンドで確認します。

```
AX620R(config)#show ipv6 route
IPv6 Routing Table - 6 entries, unlimited
Codes: C - Connected, L - Local, S - Static
       R - RIPng, O - OSPF, IA - OSPF inter area
       E1 - OSPF external type 1, E2 - OSPF external type 2, B - BGP
       s - Summary
Timers: Uptime/Age
R     2001:db8:101::/64 global [120/2]
        via fe80::c0a8:165, Tunnel0.0, 0:43:21/0:00:12
R     2001:db8:102::/64 global [120/2]
        via fe80::2a0:deff:fe65:afd1, Tunnel1.0, 0:09:42/0:00:08
C     2001:db8:103::/64 global [0/1]
        via ::, FastEthernet0/0.0, 0:46:46/0:00:00
L     2001:db8:103::/128 global [0/1]
        via ::, FastEthernet0/0.0, 0:46:47/0:00:00
L     2001:db8:103::1/128 global [0/1]
        via ::, FastEthernet0/0.0, 0:46:46/0:00:00
AX620R(config)#
```

STEP9 トンネル状態の確認

各ルーターのトンネルの状態を確認します。

■ Cisco1841:

show interfaces tunnel コマンドで確認します。

```
Cisco1841#show interfaces tunnel 0
Tunnel0 is up, line protocol is up
  Hardware is Tunnel
  MTU 1514 bytes, BW 9 Kbit, DLY 500000 usec,
     reliability 255/255, txload 1/255, rxload 1/255
  Encapsulation TUNNEL, loopback not set
  Keepalive not set
  Tunnel source 192.168.1.101 (FastEthernet0/1), destination 192.168.1.102
  Tunnel protocol/transport IPv6/IP
  Tunnel TTL 255
  Fast tunneling enabled
  Tunnel transmit bandwidth 8000 (kbps)
  Tunnel receive bandwidth 8000 (kbps)
  Last input 00:00:01, output 00:00:25, output hang never
  Last clearing of "show interface" counters never
  Input queue: 0/75/0/0 (size/max/drops/flushes); Total output drops: 13
  Queueing strategy: fifo
  Output queue: 0/0 (size/max)
```

Chapter 10 IPv6活用

```
      5 minute input rate 0 bits/sec, 0 packets/sec
      5 minute output rate 0 bits/sec, 0 packets/sec
         117 packets input, 16604 bytes, 0 no buffer
         Received 0 broadcasts, 0 runts, 0 giants, 0 throttles
         0 input errors, 0 CRC, 0 frame, 0 overrun, 0 ignored, 0 abort
         148 packets output, 15828 bytes, 0 underruns
         0 output errors, 0 collisions, 0 interface resets
         0 output buffer failures, 0 output buffers swapped out
Cisco1841#show interfaces tunnel 1
Tunnel1 is up, line protocol is up
   Hardware is Tunnel
   MTU 1514 bytes, BW 9 Kbit, DLY 500000 usec,
      reliability 255/255, txload 1/255, rxload 1/255
   Encapsulation TUNNEL, loopback not set
   Keepalive not set
   Tunnel source 192.168.1.101 (FastEthernet0/1), destination 192.168.1.103
   Tunnel protocol/transport IPv6/IP
   Tunnel TTL 255
   Fast tunneling enabled
   Tunnel transmit bandwidth 8000 (kbps)
   Tunnel receive bandwidth 8000 (kbps)
   Last input 00:00:28, output 00:00:05, output hang never
   Last clearing of "show interface" counters never
   Input queue: 0/75/0/0 (size/max/drops/flushes); Total output drops: 21
   Queueing strategy: fifo
   Output queue: 0/0 (size/max)
   5 minute input rate 0 bits/sec, 0 packets/sec
   5 minute output rate 0 bits/sec, 0 packets/sec
      129 packets input, 18828 bytes, 0 no buffer
      Received 0 broadcasts, 0 runts, 0 giants, 0 throttles
      0 input errors, 0 CRC, 0 frame, 0 overrun, 0 ignored, 0 abort
      147 packets output, 15984 bytes, 0 underruns
      0 output errors, 0 collisions, 0 interface resets
      0 output buffer failures, 0 output buffers swapped out
Cisco1841#
```

■**RTX1200:**

show status tunnel コマンドで確認します。

```
RTX1200#show status tunnel 1
TUNNEL[1]:
Description:
  Interface type: IP over IP
  Current status is Online.
  from: 2010/12/04 17:12:29
  1 minute   connection.
  Received:     (IPv4) 0 packets [0 octet]
                (IPv6) 10 packets [920 octet]
  Transmitted: (IPv4) 0 packets [0 octet]
                (IPv6) 8 packets [876 octet]
RTX1200#
RTX1200#show status tunnel 2
TUNNEL[2]:
Description:
  Interface type: IP over IP
  Current status is Online.
  from: 2010/12/04 17:12:29
  1 minute   connection.
```

```
            Received:    (IPv4) 0 packets [0 octet]
                         (IPv6) 12 packets [1264 octet]
            Transmitted: (IPv4) 0 packets [0 octet]
                         (IPv6) 8 packets [876 octet]
RTX1200#
```

■ **AX620R:**

show interfaces tunnel コマンドで確認します。

```
AX620R(config)#show interfaces Tunnel0.0
Interface Tunnel0.0 is up
  Fundamental MTU is 1480 octets
  Current bandwidth 100M b/s, QoS is disabled
  Datalink header cache type is ipv4-tunnel: 0/0 (standby/dynamic)
  IPv6 subsystem connected, physical layer is up, 0:45:40
  SNMP MIB-2:
    ifIndex is 1017
  Logical INTERFACE:
    Elapsed time after clear counters 1:32:20
    106 packets input, 11832 bytes, 0 errors
      106 unicasts, 0 non-unicasts, 0 unknown protos
      0 drops, 0 misc errors
    110 output requests, 13740 bytes, 0 errors
      110 unicasts, 0 non-unicasts
      0 overflows, 0 neighbor unreachable, 0 misc errors
    2 link-up detected, 1 link-down detected
  Encapsulation TUNNEL:
    Tunnel mode is 6-over-4
    Tunnel is ready
    Destination address is 192.168.1.101
    Source address is 192.168.1.103
    Outgoing interface is FastEthernet1/0.0
    Interface MTU is 1480
    Path MTU is 1500
    Statistics:
      106 packets input, 11832 bytes, 0 errors
      111 packets output, 13872 bytes, 0 errors
    Received ICMP messages:
      0 errors

AX620R(config)#show interfaces Tunnel1.0
Interface Tunnel1.0 is up
  Fundamental MTU is 1480 octets
  Current bandwidth 100M b/s, QoS is disabled
  Datalink header cache type is ipv4-tunnel: 0/0 (standby/dynamic)
  IPv6 subsystem connected, physical layer is up, 0:45:30
  SNMP MIB-2:
    ifIndex is 1018
  Logical INTERFACE:
    Elapsed time after clear counters 1:32:33
    84 packets input, 10808 bytes, 0 errors
      84 unicasts, 0 non-unicasts, 0 unknown protos
      0 drops, 0 misc errors
    111 output requests, 13872 bytes, 0 errors
      111 unicasts, 0 non-unicasts
      0 overflows, 0 neighbor unreachable, 0 misc errors
    2 link-up detected, 1 link-down detected
  Encapsulation TUNNEL:
    Tunnel mode is 6-over-4
```

Chapter 10 IPv6活用

```
      Tunnel is ready
      Destination address is 192.168.1.102
      Source address is 192.168.1.103
   Outgoing interface is FastEthernet1/0.0
   Interface MTU is 1480
   Path MTU is 1500
   Statistics:
      84 packets input, 10808 bytes, 0 errors
      111 packets output, 13872 bytes, 0 errors
   Received ICMP messages:
      0 errors
AX620R(config)#
```

10-2 IPv6実習サーバーの構築

IPv6 ACLなどの実習で利用するメールサーバー, DNSサーバー, WWWサーバーおよびDHCPサーバーを構築します。

今回サーバーとして利用するOSはFreeBSD8.3です。FreeBSD8.3のインストール方法はこの実習に含まれていません。必要に応じてインターネット上 (http://www.jp.freebsd.org) などから情報を収集し, インストールしてください。

本実習ではメールサーバーにPostfixを, DNSとしてBINDを用い, WWWサーバーはApache2を利用します。またDHCPサーバーとしてISC-DHCPサーバーを利用します。

■ サーバー設定情報

> **注** 今回サーバーはCisco1841のFastEthernet0/0下に設置すると想定しています。したがってデフォルトルーターは2001:db8:0:3::1です。またドメイン名はv6.example.jpとしています。
>
> DNSサーバー　　　2001:db8:0:3::53/64　　dns.v6.example.jp
> メールサーバー　　2001:db8:0:3::23/64　　mail.v6.example.jp
> WWWサーバー　　2001:db8:0:3::80/64　　www.v6.example.jp

■ 実習A インターフェースの設定

FreeBSD8.3でIPv6を利用する場合, 1つの物理インターフェースに複数の論理アドレスを割り当てることができます。本実習では上記3つのアドレスを1つの物理インターフェースに割り当てます。

STEP A-1　FreeBSD8.3インストール

可能であればcsupを用いてportsやソースを最新版にし, カーネルも再構築しておきます。最低限パッチは当てておきましょう。

なお, IPv6のみで外部と接続している場合cvsup5.jp.freebsd.orgなど一部のサーバーのみIPv6接続可能ですので注意が必要です。

STEP A-2　インターフェース設定

下記例はインターフェース名em0の場合です。インターフェース名はインストールした機器に依存しますので, **ifconifg -a** コマンドなどでインターフェース情報を確認してください。

/etc/rc.confファイルに下記を追加します。

```
ipv6_enable="YES"
ipv6_ifconfig_em0="2001:db8:0:3::53 prefixlen 64"
ipv6_ifconfig_em0_alias0="2001:db8:0:3::80 prefixlen 64"
ipv6_ifconfig_em0_alias1="2001:db8:0:3::23 prefixlen 64"
ipv6_defaultrouter="2001:db8:0:3::1"
```

STEP A-3　設定の確認

IPv6 アドレスの確認は以下の **ifconfig** コマンドで行います。

```
dns#ifconfig -a
em0: flags=8843<UP,BROADCAST,RUNNING,SIMPLEX,MULTICAST> metric 0 mtu 1500
    options=9b<RXCSUM,TXCSUM,VLAN_MTU,VLAN_HWTAGGING,VLAN_HWCSUM>
    ether 00:1c:42:e7:e9:1d
    inet6 fe80::21c:42ff:fee7:e91d%em0 prefixlen 64 scopeid 0x1
    inet6 2001:db8:0:3::53 prefixlen 64
    inet6 2001:db8:0:3::80 prefixlen 64
    inet6 2001:db8:0:3::23 prefixlen 64
    nd6 options=3<PERFORMNUD,ACCEPT_RTADV>
    media: Ethernet autoselect (1000baseT <full-duplex>)
    status: active
plip0: flags=8810<POINTOPOINT,SIMPLEX,MULTICAST> metric 0 mtu 1500
lo0: flags=8049<UP,LOOPBACK,RUNNING,MULTICAST> metric 0 mtu 16384
    options=3<RXCSUM,TXCSUM>
    inet 127.0.0.1 netmask 0xff000000
    inet6 ::1 prefixlen 128
    inet6 fe80::1%lo0 prefixlen 64 scopeid 0x3
    nd6 options=3<PERFORMNUD,ACCEPT_RTADV>
dns#
```

ルーティングテーブルの確認は以下の **netstat** コマンドで行います。

```
#netstat -r
Routing tables

Internet:
Destination        Gateway            Flags      Refs       Use     Netif Expire
default            10.0.0.1           UGS        1          7       em0
10.0.0.0           link#1             U          0          0       em0
mail               link#1             UHS        0          0       lo0
localhost          link#3             UH         0          0       lo0

Internet6:
Destination        Gateway            Flags      Netif Expire
::                 localhost          UGRS       lo0 =>
default            2001:db8:0:3::1    UGS        em0
localhost          localhost          UH         lo0
::ffff:0.0.0.0     localhost          UGRS       lo0
2001:db8:0:3::     link#1             U          em0
2001:db8:0:3::23   link#1             UHS        lo0
2001:db8:0:3::53   link#1             UHS        lo0
2001:db8:0:3::80   link#1             UHS        lo0
fe80::             localhost          UGRS       lo0
fe80::%em0         link#1             U          em0
fe80::21c:42ff:fee link#1             UHS        lo0
fe80::%lo0         link#3             U          lo0
fe80::1%lo0        link#3             UHS        lo0
ff01:1::           fe80::21c:42ff:fee U          em0
ff01:3::           localhost          U          lo0
ff02::             localhost          UGRS       lo0
ff02::%em0         fe80::21c:42ff:fee U          em0
ff02::%lo0         localhost          U          lo0
dns#
```

■実習B　DNS サーバーの導入

DNS サーバーは FreeBSD にデフォルトでインストールされていますので，機能の有効化と設定ファイルを作成します。ファイルは /etc/namedb ディレクトリに保存するものとします。

STEP B-1　BIND の有効化

/etc/rc.conf に下記設定を追加します。

```
named_enable="YES"
```

STEP B-2　制御ファイル (/etc/namedb/named.conf) 設定

以下の設定をファイルに記述します。すでに named.conf ファイルが存在する場合，外部に接続しない場合は下記内容と入れ替えるだけでも動作します。

```
options {
    directory "/etc/namedb";
    pid-file "/var/run/named/pid";
    listen-on-v6 {
       any;
          };
         };

zone "v6.example.jp" {
    type master;
    file "v6.example.jp.zone";
          };
zone "3.0.0.0.0.0.0.0.8.b.d.0.1.0.0.2.ip6.arpa"{
    type master;
    file "v6.example.jp.ipv6.rev";
                           };
zone "localhost" {
    Type master;
    file "localhost.zone";
          };
zone "0.0.0.0.0.0.0.0.0.0.0.0.0.0.0.0.0.0.0.0.0.0.0.0.0.0.0.0.0.0.0.0.ip6.arpa" {
    type master;
    file "localhost.ipv6.rev";
```

STEP B-3　IPv6 正引きファイル (/etc/namedb/v6.example.jp.zone) 設定

以下の設定をファイルに記述します。

```
$TTL 86400
@   IN SOA ns.v6.example.jp. root.v6.example.jp. (
             20110401 ; serial
             3600     ; refresh
             900      ; retry
             604800   ; expire
             86400  ) ; minimum
$ORIGIN v6.example.jp.
    IN NS    ns.
    IN MX 10 mail.
mail IN AAAA  2001:db8:0:3::23
ns   IN AAAA  2001:db8:0:3::53
www  IN AAAA  2001:db8:0:3::80
```

STEP B-4　IPv6 逆引きファイル (/etc/namedb/v6.example.jp.ipv6.rev) 設定

以下の設定をファイルに記述します。

```
$TTL 86400
  @   IN SOA ns.v6.example.jp. root.v6.example.jp. (
            20110401 ; serial
            3600     ; refresh
            900      ; retry
            604800   ; expire
            86400 )  ; minimum

                      IN NS  dns.v6.example.jp.
2.3.0.0.0.0.0.0.0.0.0.0.0.0.0.0     IN PTR mail.v6.example.jp.
5.3.0.0.0.0.0.0.0.0.0.0.0.0.0.0     IN PTR dns.v6.example.jp.
8.0.0.0.0.0.0.0.0.0.0.0.0.0.0.0     IN PTR www.v6.example.jp.
```

STEP B-5　IPv6 ローカル正引きファイル (/etc/namedb/localhost.zone) 設定

以下の設定をファイルに記述します。

```
$TTL 86400
  @   IN SOA ns.v6.example.jp. root.v6.example.jp. (
            201100401 ; serial
            3600      ; refresh
            900       ; retry
            604800    ; expire
            86400 )   ; minimum

      IN NS    ns.v6.example.jp.
      IN AAAA  ::1
```

STEP B-6　IPv6 ローカル逆引きファイル (/etc/namedb/localhost.ipv6.rev) 設定

以下の設定をファイルに記述します。

```
$TTL 86400
  @   IN SOA ns.v6.example.jp. root.v6.example.jp. (
            20110401 ; serial
            3600     ; refresh
            900      ; retry
            604800   ; expire
            86400 )  ; minimum

      IN NS    localhost.
  1   IN PTR   localhost.
```

STEP B-7　/etc/resolv.conf ファイル設定

以下の設定をファイルに記述します。

```
        domain v6.example.jp
        nameserver ::1
```

STEP B-8　BIND の起動と確認

root 権限で下記コマンドにより BIND を起動します。

```
dns#sh /etc/rc.d/named start
```

起動確認は下記コマンドを入力してください。以下のようにコマンド出力にプロセスが表示されれば起動しています。

```
dns#ps aux | grep bind
bind     1492  0.0  4.8 17444 12044  ??  Ss     7:54AM    0:00.06 /usr/sbin/named
```

STEP B-9　BIND の動作確認

次のコマンドを入力し，下記のような正引き応答を確認してください。

```
dns#dig -t AAAA www.v6.example.jp
; <<>> DiG 9.6.-ESV-R3 <<>> -t AAAA www.v6.example.jp
;; global options: +cmd
;; Got answer:
;; ->>HEADER<<- opcode: QUERY, status: NOERROR, id: 56748
;; flags: qr aa rd ra; QUERY: 1, ANSWER: 1, AUTHORITY: 1, ADDITIONAL: 0

;; QUESTION SECTION:
;www.v6.example.jp.             IN      AAAA

;; ANSWER SECTION:
www.v6.example.jp.      86400   IN      AAAA    2001:db8:0:3::80

;; AUTHORITY SECTION:
v6.example.jp.          86400   IN      NS      dns.

;; Query time: 20 msec
;; SERVER: ::1#53(::1)
;; WHEN: Mon Jun 27 18:34:51 2011
;; MSG SIZE  rcvd: 80
```

次に逆引きとして次のコマンドを入力して応答を確認してください。

```
dns#dig -x 2001:db8:0:3::53

; <<>> DiG 9.6.-ESV-R3 <<>> -x 2001:db8:0:3::53
;; global options: +cmd
;; Got answer:
;; ->>HEADER<<- opcode: QUERY, status: NOERROR, id: 43052
;; flags: qr aa rd ra; QUERY: 1, ANSWER: 1, AUTHORITY: 1, ADDITIONAL: 1

;; QUESTION SECTION:
;3.5.0.0.0.0.0.0.0.0.0.0.0.0.0.0.3.0.0.0.0.0.0.0.8.b.d.0.1.0.0.2.ip6.arpa. IN PTR

;; ANSWER SECTION:
3.5.0.0.0.0.0.0.0.0.0.0.0.0.0.0.3.0.0.0.0.0.0.0.8.b.d.0.1.0.0.2.ip6.arpa.
86400IN PTR dns.v6.example.jp.

;; AUTHORITY SECTION:
3.0.0.0.0.0.0.0.8.b.d.0.1.0.0.2.ip6.arpa. 86400 IN NS dns.v6.example.jp.

;; ADDITIONAL SECTION:
dns.v6.example.jp.      86400   IN      AAAA    2001:db8:0:3::53

;; Query time: 0 msec
```

```
;; SERVER: ::1#53(::1)
;; WHEN: Mon Jun 27 18:35:06 2011
;; MSG SIZE  rcvd: 163
```

■実習 C　Web サーバーの導入

Apache2 をサーバーに導入します。導入には ports を利用します。

STEP C-1　ports からのインストール

root 権限で下記コマンドを実行してください。

```
#cd /usr/ports/www/apache22
#make
```

Apache 導入時のオプションを尋ねられますが，IPv6 機能を有効にしてください。他の機能は特に必要ありません。また perl のオプションについても必須機能はこの実習にはありません。m4 や libiconv，python などについても IPv6 以外のオプション設定は不要で，デフォルトでインストールしてください。make 終了後，以下のコマンドを実行してインストールは完了です。

```
#make install
```

STEP C-2　/usr/local/etc/apache22/httpd.conf ファイルの設定

以下の設定をファイルに記述します。

```
Listen 80
User www
Group www
ServerAdmin admin@v6.example.jp
DocumentRoot "/usr/local/www/apache22/data"
ErrorLog /var/log/http-error.log
```

STEP C-3　Apache の起動

```
#/usr/local/sbin/apachectl start
```

STEP C-4　Apache の動作確認

PC から Web ブラウザを用いて動作を確認します。

```
http://[2001:db8:0:3::80]
```
または
```
http://www.v6.example.jp
```

■実習 D　メールサーバーの導入

Postfix をサーバーに導入します。導入には ports を利用します。

STEP D-1　ports からのインストール

root 権限で下記コマンドを実行してください。

10-2 IPv6実習サーバーの構築

```
#cd /usr/ports/mail/postfix
#make
```

オプションの設定は不要です。コンパイル終了後，下記コマンドでインストールします。インストール時に postfix 運用に必要なユーザー登録を確認されますが，指定のとおりとしてください。

```
#make install
```

また postfix を有効化するか尋ねられますので有効化してください。

STEP D-2 /usr/local/etc/postfix/mail.cf ファイルの設定

以下の設定をファイルに追加記述します。

```
myhostname=mail.v6.example.jp
mydomain=v6.example.jp
myorigin=$mydomain
inet_interface=all
mydestination=$myhostanme, localhost.$mydomain, localhost. $mydomain
mynetworks=[2001:db8::]/32,[::1]/128
alias_maps=hash:/etc/aliases
alias_database=hash:/etc/aliases
mail_spool_directory=/var/mail
smtpd_banner=$myhostanme ESMTP unknown
inet_protocols=ipv6
```

STEP D-3 sendmail の停止

```
#cd /etc/rc.d
#rm -rf sendmail
```

STEP D-4 Postfix の起動

```
#/usr/local/sbin/postfix start
```

STEP D-5 プロセス動作確認

以下の入力で，プロセスが動作しているか確認してください。

```
dns#ps aux | grep postfix
postfix  1488  0.0  0.6  3500  1564  ??  I    6:48PM  0:00.00 qmgr -l -t fifo
postfix  1489  0.0  0.6  3500  1540  ??  I    6:48PM  0:00.00 pickup -l -t fi
```

■実習 E DHCP サーバーの導入

FreeBSD の ports から ISC-DHCP サーバーをインストールします。

STEP E-1 ports からのインストール

root 権限で下記コマンドを実行してください。オプションはデフォルト選択とします。

```
#cd /usr/ports/net/isc-dhcp41-server
#make
```

```
#make install
```

STEP E-2　作業ファイルの作成

インストール後，下記コマンドを実行し，作業ファイル /var/db/dhcpd6.lease を作成してください。

```
dns#touch /var/db/dhcpd6.lease
```

STEP E-3　/etc/dhcpd.conf ファイルの作成

以下の設定をファイルに記述します。

```
default-lease-time 600;
max-lease-time 7200;
log-facility local7;
subnet6 2001:db8:0:3::/64 {
    range6 2001:db8:0:3::129 2001:db8:0:3::254;
    option dhcp6.name-servers 2001:db8:0:3::53;
    option dhcp6.domain-search "v6.example.jp";
    }
```

STEP E-4　DHCP サーバーの起動

root 権限で下記コマンドにより DHCP サーバーを起動します。

```
#/usr/local/sbin/dhcpd -6 em0
```

以下のような応答があれば動作していることが確認できます。

```
Internet Systems Consortium DHCP Server 4.1.1-P1
Copyright 2004-2010 Internet Systems Consortium.
All rights reserved.
For info, please visit https://www.isc.org/software/dhcp/
Wrote 0 leases to leases file.
Bound to *:547
Listening on Socket/5/em0/2001:db8:0:3::/64
Sending on   Socket/5/em0/2001:db8:0:3::/64
```

STEP E-5　DHCP サーバー起動時自動起動

サーバー起動時に自動的に DHCP サーバーを起動させるには，/usr/local/etc/rc.d/dhcpd.sh ファイルに下記内容を作成しておきます。

```
#!/bin/sh
if [ -f /usr/local/sbin/dhcpd -a -f /usr/local/etc/dhcpd.conf ]; then
    /usr/local/sbin/dhcpd -6 em0
fi
```

Chapter 11

IPv6 セキュリティ

11-1 IPv6 ACL
11-2 IPv6でのIPsec

Chapter 11　IPv6セキュリティ

11-1　IPv6 ACL

ここでは下記ネットワークに接続されたサーバーへのアクセスをコントロールします。

この実習は第4章"スタティックルートの設定"を完了し，10-2節"IPv6実習サーバーの構築"も完了させた後で行ってください。Cisco1841のFastEthernet 0/0下にサーバーを設置し，RTX1200とAX620Rの下にPCを設置した状況から開始してください。

この節のトポロジーは以下のとおりです。

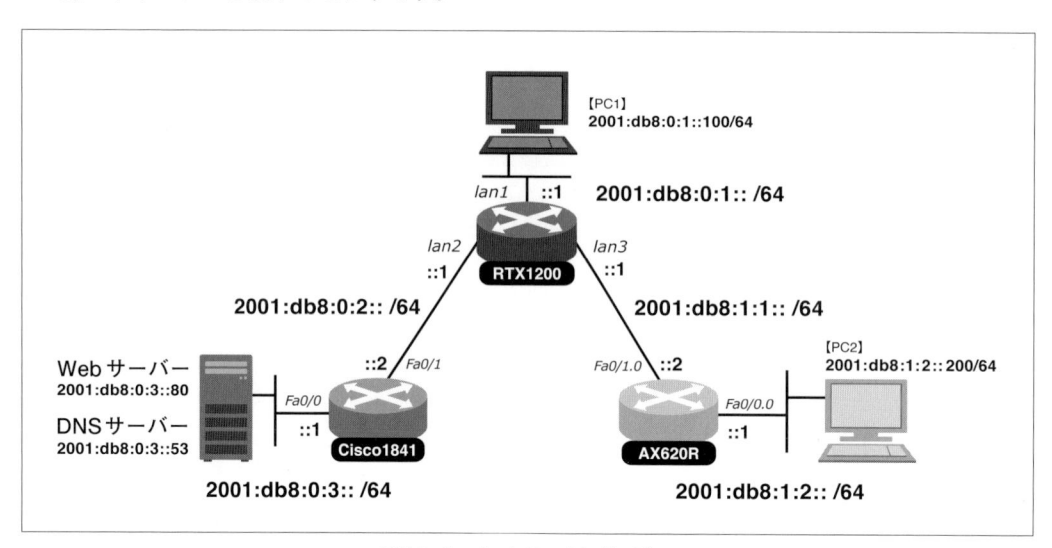

図11-1　ネットワークトポロジー

STEP1　ACL設定前の状況確認

RTX1200のlan1側，AX620RのFa0/0.0側にそれぞれクライアントPCを設置します。

RTX1200下クライアントPC1アドレスは2001:db8:0:1::100/64，AX620R下クライアントPC2のアドレスは2001:db8:1:2::200/64とし，DNSサーバーは2001:db8:0:3::53を設定してください。またデフォルトルートは上記トポロジーをもとに，もよりのルーターを設定してください。

2台のクライアントでサーバーに `ping` を実行し，通信確認を行ってください。

```
ping 2001:db8:0:3::53
```

応答があることを確認してください。

```
C:\Users>ping 2001:db8:0:3::53
2001:db8:0:3::53 に ping を送信しています 32 バイトのデータ:
2001:db8:0:3::53 からの応答: 時間 <1ms
2001:db8:0:3::53 からの応答: 時間 =1ms
2001:db8:0:3::53 からの応答: 時間 <1ms
2001:db8:0:3::53 からの応答: 時間 =1ms

2001:db8:0:3::53 の ping 統計:
    パケット数: 送信 = 4, 受信 = 4, 損失 = 0 (0% の損失),
ラウンド トリップの概算時間 (ミリ秒):
    最小 = 0ms, 最大 = 1ms, 平均 = 0ms
```

次にDNSの問い合わせを行います。

```
nslookup www.v6.example.jp
```

以上の入力で回答 (2001:db8:0:3::80) があることを確認してください。

```
C:\Users>nslookup www.v6.example.jp
サーバー:  dns.v6.example.jp
Address:  2001:db8:0:3::53

名前:     www.v6.example.jp
Address:  2001:db8:0:3::80
```

最後にクライアントからブラウザーを起動し，ホームページを表示させてください。

http://www.v6.example.jp

STEP2 ACL 設定

今回設定する ACL は Cisco1841 のインターフェース FastEthernet 0/1 から入ってくるパケットを対象とします。また ACL 名は acl-work1 とします。

ACL 設定内容は以下のとおりです。

> 任意の ipv6 アドレスからの DNS 参照は許可。
> PC1 が所属するネットワークから WWW 閲覧を禁止。
> PC2 が所属するネットワークから WWW 閲覧を許可。
> **ping** 応答は禁止。

Cisco1841 に下記設定を入力します。

```
Cisco1841(config)#ipv6 access-list acl-work1
Cisco1841(config-ipv6-acl)#permit udp any host 2001:db8:0:3::53 eq domain
Cisco1841(config-ipv6-acl)#deny tcp 2001:db8:0:1::/64 host 2001:db8:0:3::80 eq www
Cisco1841(config-ipv6-acl)#permit tcp 2001:db8:1:2::/64 host 2001:db8:0:3::80 eq www
Cisco1841(config-ipv6-acl)#deny icmp any any
Cisco1841(config-ipv6-acl)#exit
Cisco1841(config)#interface FastEthernet0/1
Cisco1841(config-if)#ipv6 traffic-filter acl-work1 in
Cisco1841(config-if)#exit
```

STEP3 ACL 設定確認

Cisco1841 に設定したアクセスリストを **show access-lists** コマンドで確認してください。

```
Cisco1841#show access-lists
IPv6 access list acl-work1
    permit udp any host 2001:DB8:0:3::53 eq domain sequence 10
    deny tcp 2001:DB8:0:1::/64 host 2001:DB8:0:3::80 eq www sequence 20
    permit tcp 2001:DB8:1:2::/64 host 2001:DB8:0:3::80 eq www sequence 30
    deny icmp any any sequence 40
Cisco1841#
```

STEP1 で実施した内容を再度実行し，違いを比較し想定したコントロールができているか確認してください。

再度 **show access-lists** コマンドを実行してください。どのアクセスリストが何回該当したかのデータ括弧内に表示されます。

Chapter 11 IPv6セキュリティ

```
Cisco1841#show access-lists
IPv6 access list acl-work1
    permit udp any host 2001:DB8:0:3::53 eq domain(16 matches) sequence 10
    deny tcp 2001:DB8:0:1::/64 host 2001:DB8:0:3::80 eq www(8 matches) sequence 20
    permit tcp 2001:DB8:1:2::/64 host 2001:DB8:0:3::80 eq www (8 matches)sequence 30
    deny icmp any any (17 matches) sequence 40
Cisco1841#
```

■ Cisco1841 最終設定

```
Cisco1841#show run
Building configuration...

Current configuration : 1410 bytes
!
version 12.4
service timestamps debug datetime msec
service timestamps log datetime msec
no service password-encryption
!
hostname Cisco1841
!
boot-start-marker
boot-end-marker
!
no aaa new-model
!
dot11 syslog
!
ip cef
!
ipv6 unicast-routing
multilink bundle-name authenticated
!
archive
 log config
  hidekeys
!
interface FastEthernet0/0
 no ip address
 duplex auto
 speed auto
 ipv6 address 2001:DB8:0:3::1/64
 ipv6 enable
!
interface FastEthernet0/1
 no ip address
 duplex auto
 speed auto
 ipv6 address 2001:DB8:0:2::2/64
 ipv6 enable
 ipv6 traffic-filter acl-work1 in
!
ip forward-protocol nd
!
no ip http server
no ip http secure-server
!
ipv6 route 2001:DB8:0:1::/64 2001:DB8:0:2::1
```

```
ipv6 route 2001:DB8:1:1::/64 2001:DB8:0:2::1
ipv6 route 2001:DB8:1:2::/64 2001:DB8:0:2::1
!
ipv6 access-list acl-work1
 permit udp any host 2001:DB8:0:3::53 eq domain
 deny tcp 2001:DB8:0:1::/64 host 2001:DB8:0:3::80 eq www
 permit tcp 2001:DB8:1:2::/64 host 2001:DB8:0:3::80 eq www
 deny icmp any any
!
control-plane
!
line con 0
line aux 0
line vty 0 4
 login
!
end
```

Chapter 11 IPv6セキュリティ

11-2 IPv6でのIPsec

この実習では，WAN接続をIPsecトンネルとし，ネットワーク資源にセキュアにアクセスする方法を学習します。

3台のルーターが接続されています。RTX1200のLAN側(lan1)にサーバーが設置されており，Cisco1841(Fa0/0)とAX620R(Fa0/0.0)のLAN側にクライアントが存在するとします。

なお，この実習は第4章"スタティックルートの設定"の4-1節STEP4を完了した状況から開始してください。

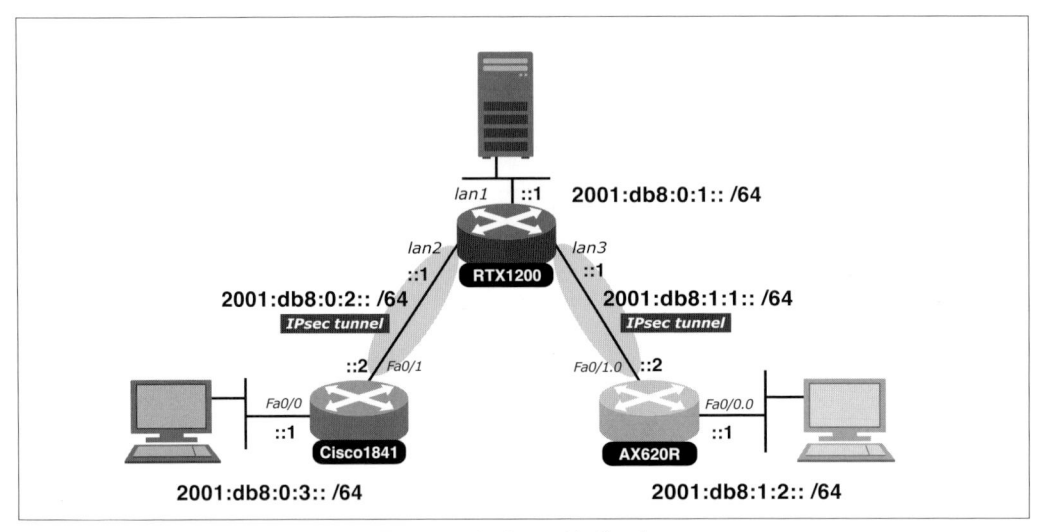

図11-2 ネットワークトポロジー

アドレスの割当は以下のとおりです。

表11-1 インターフェースアドレス割当一覧

ルーター		インターフェース	IPv6アドレス
Cisco1841	LAN側	FastEthernet 0/0 (Fa0/0)	2001:db8:0:3::1/64
	WAN側	FastEthernet 0/1 (Fa0/1)	2001:db8:0:2::2/64
RTX1200	LAN側	lan1	2001:db8:0:1::1/64
	WAN側	lan2	2001:db8:0:2::1/64
	WAN側	lan3	2001:db8:1:1::1/64
AX620R	LAN側	FastEthernet 0/0.0 (Fa0/0.0)	2001:db8:1:2::1/64
	WAN側	FastEthernet 0/1.0 (Fa0/1.0)	2001:db8:1:1::2/64

今回のIPsec接続実習で設定するパラメータは以下のとおりです。それぞれのパラメータに関する説明はここでは省略します。

11-2 IPv6でのIPsec

表11-2　RTX1200とCisco1841間IPsecパラメーター覧

	RTX1200	Cisco1841
WAN側アドレス	2001:db8:0:2::1/64	2001:db8:0:2::2/64
LAN側アドレス	2001:db8:0:1::1/64	2001:db8:0:3::1/64
IKE認証形式	pre-shared	pre-shared
IKE認証共通キー	KEY1	KEY1
Diffie-Hellmanグループ	1024bit	1024bit
IKE暗号化方式	AES-128	AES-128
IKE認証ハッシュアルゴリズム	SHA1	SHA1
SAライフタイム	43200秒	43200秒
SAモード	トンネル	トンネル
セキュリティプロトコル	ESP	ESP
暗号方式	AES-128	AES-128
認証用ハッシュアルゴリズム	SHA1	SHA1

表11-3　RTX1200とAX620R間IPsecパラメーター覧

	RTX1200	AX620R
WAN側アドレス	2001:db8:1:1::1/64	2001:db8:1:1::2/64
LAN側アドレス	2001:db8:0:1::1/64	2001:db8:1:2::1/64
IKE認証形式	pre-shared	pre-shared
IKE認証共通キー	KEY2	KEY2
Diffie-Hellmanグループ	1024bit	1024bit
IKE暗号化方式	AES-128	AES-128
IKE認証ハッシュアルゴリズム	SHA1	SHA1
SAライフタイム	43200秒	43200秒
SAモード	トンネル	トンネル
セキュリティプロトコル	ESP	ESP
暗号方式	AES-128	AES-128
認証用ハッシュアルゴリズム	SHA1	SHA1

STEP1　事前準備

4-1節"マルチベンダー機器による実習"のSTEP4まで終了していることを確認します。それぞれのルーターから隣接のルーターに **ping** を実行し，他のネットワークと通信ができることを確認してください。

■ Cisco1841：

```
Cisco1841#ping 2001:db8:0:2::1

Type escape sequence to abort.
Sending 5, 100-byte ICMP Echos to 2001:DB8:0:2::1, timeout is 2 seconds:
!!!!!
Success rate is 100 percent (5/5), round-trip min/avg/max = 0/0/0 ms
```

■ RTX1200：

```
RTX1200#ping6 2001:db8:0:2::2
received from 2001:db8:0:2::2, icmp_seq=0 hlim=64 time=0.449ms
received from 2001:db8:0:2::2, icmp_seq=1 hlim=64 time=0.417ms
received from 2001:db8:0:2::2, icmp_seq=2 hlim=64 time=0.414ms
received from 2001:db8:0:2::2, icmp_seq=3 hlim=64 time=0.417ms
received from 2001:db8:0:2::2, icmp_seq=4 hlim=64 time=0.415ms
```

Chapter 11 IPv6セキュリティ

```
received from 2001:db8:0:2::2, icmp_seq=5 hlim=64 time=0.417ms

6 packets transmitted, 6 packets received, 0.0% packet loss
round-trip min/avg/max = 0.414/0.421/0.449 ms

RTX1200#ping6 2001:db8:1:1::2
received from 2001:db8:1:1::2, icmp_seq=0 hlim=64 time=0.449ms
received from 2001:db8:1:1::2, icmp_seq=1 hlim=64 time=0.417ms
received from 2001:db8:1:1::2, icmp_seq=2 hlim=64 time=0.414ms
received from 2001:db8:1:1::2, icmp_seq=3 hlim=64 time=0.417ms
received from 2001:db8:1:1::2, icmp_seq=4 hlim=64 time=0.415ms
received from 2001:db8:1:1::2, icmp_seq=5 hlim=64 time=0.417ms

6 packets transmitted, 6 packets received, 0.0% packet loss
round-trip min/avg/max = 0.414/0.421/0.449 ms
```

■ **AX620R：**

```
AX620R(config)#ping6 2001:db8:1:1::1
PING 2001:db8:1:1::2 > 2001:db8:1:1::1 56 data bytes
64 bytes from 2001:db8:1:1::1 icmp_seq=0 hlim=64 time=0.455 ms
64 bytes from 2001:db8:1:1::1 icmp_seq=1 hlim=64 time=0.418 ms
64 bytes from 2001:db8:1:1::1 icmp_seq=2 hlim=64 time=0.414 ms
64 bytes from 2001:db8:1:1::1 icmp_seq=3 hlim=64 time=0.412 ms
64 bytes from 2001:db8:1:1::1 icmp_seq=4 hlim=64 time=0.420 ms

--- 2001:db8:1:1::1 ping statistics ---
5 packets transmitted, 5 packets received, 0% packet loss
round-trip (ms)   min/avg/max = 0.412/0.423/0.455
```

もし `ping` が成功しない場合，第4章"スタテックルートの設定"を参照し，トラブルシューティングを行ってから次の STEP に進んでください。

STEP2-1 Cisco1841 と RTX1200 間 IPsec 設定

Cisco1841 と RTX1200 間 IPsec 設定を行います。

■ **Cisco1841：**

IKE のポリシーの優先度を 10 とし，aes-128/sha1/1024 ビット，lifetime43200 秒の設定を入力します。

```
Cisco1841(config)#crypto isakmp policy 10
Cisco1841(config-isakmp)#encryption aes
Cisco1841(config-isakmp)#hash sha
Cisco1841(config-isakmp)#authentication pre-share
Cisco1841(config-isakmp)#group 2
Cisco1841(config-isakmp)#lifetime 43200
Cisco1841(config-isakmp)#exit
```

対向ルーターアドレス 2001:db8:0:2::1 に対して事前交換設定キー (KEY1) を設定します。

```
Cisco1841(config)#crypto isakmp key 6 KEY1 address ipv6 2001:db8:0:2::1/64
```

IPsec のトランスフォームセット (test1) を esp-aes，esp-sha1 として設定します。

```
Cisco1841(config)#crypto ipsec transform-set test1 esp-aes esp-sha-hmac
Cisco1841(cfg-crypto-trans)#exit
```

このトランスフォームセットを IPsec プロファイル名 yamaha として割り当てます。

11-2 IPv6でのIPsec

```
Cisco1841(config)#crypto ipsec profile yamaha
Cisco1841(ipsec-profile)#set transform-set test1
```

このIPsecプロファイルをトンネルインターフェース0に割り当てます。

```
Cisco1841(config)#interface Tunnel0
Cisco1841(config-if)#no ip address
Cisco1841(config-if)#ipv6 unnumbered FastEthernet0/1
Cisco1841(config-if)#tunnel source FastEthernet0/1
Cisco1841(config-if)#tunnel destination 2001:db8:0:2::1
Cisco1841(config-if)#tunnel mode ipsec ipv6
Cisco1841(config-if)#tunnel protection ipsec profile yamaha
Cisco1841(config-if)#exit
```

新たにこのインターフェースを利用する経路を設定します。

```
Cisco1841(config)#ipv6 route 2001:db8:0:1::/64 Tunnel0
```

■ RTX1200：

トンネルインターフェース1にポリシー100として，aes-128/sha1/1024ビット/lifetime43200秒の設定，および事前交換設定キー（KEY1）を投入します。

```
RTX1200#tunnel select 1
RTX1200tunnel1#tunnel encapsulation ipsec
RTX1200tunnel1#ipsec tunnel 100
RTX1200tunnel1#ipsec sa policy 100 1 esp aes-cbc sha-hmac
RTX1200tunnel1#ipsec ike encryption 1 aes-cbc
RTX1200tunnel1#ipsec ike duration ike-sa 1 43200
RTX1200tunnel1#ipsec ike group 1 modp1024
RTX1200tunnel1#ipsec ike local address 1 2001:db8:0:2::1
RTX1200tunnel1#ipsec ike local id 1 ::/0
RTX1200tunnel1#ipsec ike pre-shared-key 1 text KEY1
RTX1200tunnel1#ipsec ike remote address 1 2001:db8:0:2::2
RTX1200tunnel1#ipsec ike remote id 1 ::/0
RTX1200tunnel1#tunnel enable 1
RTX1200tunnel1#ipsec auto refresh on
```

新たにこのインターフェースを利用する経路を設定します。

```
RTX1200tunnel1#ipv6 route 2001:db8:0:3::/64 gateway tunnel 1
```

STEP2-2 Cisco1841とRTX1200間IPsec確認

pingおよびSAの状態を確認します。

■ Cisco1841：

```
Cisco1841#ping 2001:db8:0:1::1

Type escape sequence to abort.
Sending 5, 100-byte ICMP Echos to 2001:DB8:0:1::1, timeout is 2 seconds:
!!!!!
Success rate is 100 percent (5/5), round-trip min/avg/max = 0/0/4 ms


Cisco1841#show crypto ipsec sa

interface: Tunnel0
    Crypto map tag: Tunnel0-head-0, local addr 2001:DB8:0:2::2

  protected vrf: (none)
```

Chapter 11 IPv6セキュリティ

```
      local  ident (addr/mask/prot/port): (::/0/0/0)
      remote ident (addr/mask/prot/port): (::/0/0/0)
      current_peer 2001:DB8:0:2::1 port 500
        PERMIT, flags={origin_is_acl,}
       #pkts encaps: 19, #pkts encrypt: 19, #pkts digest: 19
       #pkts decaps: 0, #pkts decrypt: 0, #pkts verify: 0
       #pkts compressed: 0, #pkts decompressed: 0
       #pkts not compressed: 0, #pkts compr. failed: 0
       #pkts not decompressed: 0, #pkts decompress failed: 0
       #send errors 0, #recv errors 0

        local crypto endpt.: 2001:DB8:0:2::2,
        remote crypto endpt.: 2001:DB8:0:2::1
        path mtu 1460, ip mtu 1460, ip mtu idb Tunnel0
        current outbound spi: 0x2BA87830(732461104)
        PFS (Y/N): N, DH group: none

        inbound esp sas:
       spi: 0xA33D5694(2738706068)
          transform: esp-aes esp-sha-hmac ,
          in use settings ={Tunnel, }
          conn id: 2001, flow_id: FPGA:1, sibling_flags 80000046, crypto map: Tunnel0-head-0
          sa timing: remaining key lifetime (k/sec): (4377706/3530)
          IV size: 16 bytes
          replay detection support: Y
          Status: ACTIVE

        inbound ah sas:

        inbound pcp sas:

        outbound esp sas:
         spi: 0x2BA87830(732461104)
          transform: esp-aes esp-sha-hmac ,
          in use settings ={Tunnel, }
          conn id: 2002, flow_id: FPGA:2, sibling_flags 80000046, crypto map: Tunnel0-head-0
          sa timing: remaining key lifetime (k/sec): (4377703/3530)
          IV size: 16 bytes
          replay detection support: Y
       Status: ACTIVE

        outbound ah sas:

        outbound pcp sas:

Cisco1841#show crypto isakmp sa
IPv4 Crypto ISAKMP SA
dst             src             state           conn-id status

IPv6 Crypto ISAKMP SA

 dst: 2001:DB8:0:2::2
 src: 2001:DB8:0:2::1
 state: QM_IDLE          conn-id:   1002 status: ACTIVE

Cisco1841#show crypto isakmp sa
IPv4 Crypto ISAKMP SA
dst             src             state           conn-id status

IPv6 Crypto ISAKMP SA
```

```
   dst: 2001:DB8:0:2::2
   src: 2001:DB8:0:2::1
   state: QM_IDLE          conn-id:    1002 status: ACTIVE


Cisco1841#show ipv6 route
IPv6 Routing Table - Default - 7 entries
Codes: C - Connected, L - Local, S - Static, U - Per-user Static route
       B - BGP, M - MIPv6, R - RIP, I1 - ISIS L1
       I2 - ISIS L2, IA - ISIS interarea, IS - ISIS summary, D - EIGRP
       EX - EIGRP external
       O - OSPF Intra, OI - OSPF Inter, OE1 - OSPF ext 1, OE2 - OSPF ext 2
       ON1 - OSPF NSSA ext 1, ON2 - OSPF NSSA ext 2
S   2001:DB8:0:1::/64 [1/0]
     via 2001:DB8:0:2::1
     via Tunnel0, directly connected
C   2001:DB8:0:2::/64 [0/0]
     via FastEthernet0/1, directly connected
L   2001:DB8:0:2::2/128 [0/0]
     via FastEthernet0/1, receive
C   2001:DB8:0:3::/64 [0/0]
     via FastEthernet0/0, directly connected
L   2001:DB8:0:3::1/128 [0/0]
     via FastEthernet0/0, receive
S   2001:DB8:1:1::/64 [1/0]
     via 2001:DB8:0:2::1
L   FF00::/8 [0/0]
     via Null0, receive
Cisco1841#
```

■ RTX1200：

```
RTX1200#ping6 2001:db8:0:3::1
received from 2001:db8:0:3::1, icmp_seq=0 hlim=64 time=0.449ms
received from 2001:db8:0:3::1, icmp_seq=1 hlim=64 time=0.417ms
received from 2001:db8:0:3::1, icmp_seq=2 hlim=64 time=0.414ms
received from 2001:db8:0:3::1, icmp_seq=3 hlim=64 time=0.417ms
received from 2001:db8:0:3::1, icmp_seq=4 hlim=64 time=0.415ms
received from 2001:db8:0:3::1, icmp_seq=5 hlim=64 time=0.417ms

6 packets transmitted, 6 packets received, 0.0% packet loss
round-trip min/avg/max = 0.414/0.421/0.449 ms
RTX1200#

RTX1200#show ipsec sa
Total: isakmp:1 send:1 recv:1

sa   sgw isakmp connection   dir  life[s] remote-id
-----------------------------------------------------------------------
1    1    -    isakmp         -    28664  2001:db8:0:2::2
2    1    -    tun[001]esp   send  28630  2001:db8:0:2::2
3    1    -    tun[001]esp   recv  28630  2001:db8:0:2::2

RTX1200#show ipv6 route
Destination              Gateway          Interface  Type
2001:db8:0:1::/64        -                LAN1       implicit
2001:db8:0:2::/64        -                LAN2       implicit
2001:db8:0:3::/64        -                TUNNEL[1]  static
2001:db8:1:1::/64        -                LAN3       implicit
RTX1200tunnel1#
```

Chapter 11 IPv6セキュリティ

STEP2-3 AX620RとRTX1200間IPsec設定

■ **AX620R:**

IPsecを有効にする通信の設定のためアクセスリスト(list1)を設定します。

```
AX620R(config)#ipv6 access-list list1 permit ip src any dest any
```

ikeポリシー(ike-prop)としてaes-128/sha1/1024ビット/lifetime43200秒の設定を入力します。

```
AX620R(config)#ike proposal ike-prop encryption aes hash sha group 1024-bit lifetime 43200
```

対向ルーターアドレス2001:db8:1:1::1に対して事前交換設定キー(KEY2)を設定します。

```
AX620R(config)#ike policy ike-policy peer 2001:db8:1:1::1 key KEY2 ike-prop
```

IPsecポリシー(ipsec-group)としてesp-aes, esp-sha1として設定します。

```
AX620R(config)#ipsec autokey-proposal ipsec-group esp-aes esp-sha
```

IPsecの設定を行います。

```
AX620R(config)#ipsec autokey-map ipsec-policy list1 peer 2001:db8:1:1::1 ipsec-group
AX620R(config)#ipsec local-id ipsec-policy 2001:db8:1:2::/64
AX620R(config)#ipsec remote-id ipsec-policy 2001:db8:0:1::/64
```

インターフェース(tunnel0/0.0)を設定します。

```
AX620R(config)#interface Tunnel0.0
AX620R(config-if)#tunnel mode ipsec
AX620R(config-if)#no ip address
AX620R(config-if)#ipv6 enable
AX620R(config-if)#ipv6 unnumbered FastEthernet0/0.0
AX620R(config-if)#ipsec policy tunnel ipsec-policy out
AX620R(config-if)#no shutdown
AX620R(config-if)#exit
```

RTX1200との経路を設定します。

```
AX620R(config)#ipv6 route 2001:db8:0:1::/64 tunnel0.0
```

■ **RTX1200:**

トンネルインターフェース2にポリシー200として、aes-128/sha1/1024ビット/lifetime43200秒の設定、および事前交換設定キー(KEY2)を投入します。

```
RTX1200#tunnel select 2
RTX1200tunnel2#tunnel encapsulation ipsec
RTX1200tunnel2#ipsec tunnel 200
RTX1200tunnel2#ipsec sa policy 200 2 esp aes-cbc sha-hmac
RTX1200tunnel2#ipsec ike encryption 2 aes-cbc
RTX1200tunnel2#ipsec ike duration ike-sa 2 43200
RTX1200tunnel2#ipsec ike group 2 modp1024
RTX1200tunnel2#ipsec ike local address 2 2001:db8:1:1::1
RTX1200tunnel2#ipsec ike local id 2 ::/0
RTX1200tunnel2#ipsec ike pre-shared-key 2 text KEY2
RTX1200tunnel2#ipsec ike remote address 2 2001:db8:1:1::2
RTX1200tunnel2#ipsec ike remote id 2 ::/0
RTX1200tunnel2#tunnel enable 2
RTX1200tunnel2#ipsec auto refresh on
```

新たにこのインターフェースを利用する経路を設定します。

```
#ipv6 route 2001:db8:1:2::/64 gateway tunnel 2
```

11-2 IPv6でのIPsec

STEP2-4　AX620RとRTX1200間 IPsec 確認

ping および SA の状態を確認します。

■ **AX620R：**

```
AX620R(config)#ping6 2001:db8:0:1::1
PING 2001:db8:1:2::1 > 2001:db8:0:1::1 56 data bytes
64 bytes from 2001:db8:0:1::1 icmp_seq=1 hlim=64 time=0.849 ms
64 bytes from 2001:db8:0:1::1 icmp_seq=2 hlim=64 time=0.745 ms
64 bytes from 2001:db8:0:1::1 icmp_seq=3 hlim=64 time=0.764 ms
64 bytes from 2001:db8:0:1::1 icmp_seq=4 hlim=64 time=0.741 ms

--- 2001:db8:0:1::1 ping statistics ---
5 packets transmitted, 4 packets received, 20% packet loss
round-trip (ms)  min/avg/max = 0.741/0.774/0.849
AX620R(config)#
AX620R(config)#show ike sa
ISAKMP SA - 1 configured, 1 created
Local address is 2001:db8:1:1::2, port is 500
Remote address is 2001:db8:1:1::1, port is 500
  IKE policy name is ike-policy
  Direction is responder
  Initiator's cookie is 0x38abac1b97821b2d
  Responder's cookie is 0x2290ea92696ce05e
  Exchange type is main mode
  State is established
  Authentication method is pre-shared
  Encryption algorithm is aes-128
  Hash algorithm is sha1
  DH group is modp1024, lifetime is 28656 seconds
  #ph1 success: 1, #ph1 failure: 0
  #ph1 hash err: 0, #ph1 timeout: 0, #ph1 resend: 0
  #ph2 success: 1, #ph2 failure: 11
  #ph2 hash err: 0, #ph2 timeout: 0, #ph2 resend: 0

AX620R(config)#show ipsec sa
IPsec SA - 1 configured, 2 created
Interface is Tunnel0.0
  Key policy map name is ipsec-policy
    Tunnel mode, 6-over-6, autokey-map
    Local address is 2001:db8:1:1::2
    Remote address is 2001:db8:1:1::1
    Outgoing interface is FastEthernet0/1.0
    Interface MTU is 1422, path MTU is 1500
    Inbound:
      ESP, SPI is 0xf8df0589(4175365513)
        Transform is ESP-AES-128-HMAC-SHA-96
        Remaining lifetime is 28778 seconds
      Replay detection support is on
    Outbound:
      ESP, SPI is 0x7245f18b(1917186443)
        Transform is ESP-AES-128-HMAC-SHA-96
        Remaining lifetime is 28778 seconds
      Replay detection support is on
    Perfect forward secrecy is off

AX620R(config)#show ipv6 route
IPv6 Routing Table - 8 entries, unlimited
Codes: C - Connected, L - Local, S - Static
```

Chapter 11 IPv6セキュリティ

```
        R - RIPng, O - OSPF, IA - OSPF inter area
        E1 - OSPF external type 1, E2 - OSPF external type 2, B - BGP
        s - Summary
Timers: Uptime/Age
S       2001:db8:0:1::/64 global [1/1]
            via ::, Tunnel0.0, 0:00:46/0:00:00
S       2001:db8:0:2::/64 global [1/1]
            via 2001:db8:1:1::1, FastEthernet0/1.0, 0:43:05/0:00:00
C       2001:db8:1:1::/64 global [0/1]
            via ::, FastEthernet0/1.0, 0:45:12/0:00:00
L       2001:db8:1:1::/128 global [0/1]
            via ::, FastEthernet0/1.0, 0:45:13/0:00:00
L       2001:db8:1:1::2/128 global [0/1]
            via ::, FastEthernet0/1.0, 0:45:12/0:00:00
C       2001:db8:1:2::/64 global [0/1]
            via ::, FastEthernet0/0.0, 0:45:13/0:00:00
L       2001:db8:1:2::/128 global [0/1]
            via ::, FastEthernet0/0.0, 0:45:14/0:00:00
L       2001:db8:1:2::1/128 global [0/1]
            via ::, FastEthernet0/0.0, 0:45:13/0:00:00
```

■ RTX1200：

```
RTX1200#ping6 2001:db8:1:2::1
received from 2001:db8:1:2::1, icmp_seq=0 hlim=64 time=0.449ms
received from 2001:db8:1:2::1, icmp_seq=1 hlim=64 time=0.417ms
received from 2001:db8:1:2::1, icmp_seq=2 hlim=64 time=0.414ms
received from 2001:db8:1:2::1, icmp_seq=3 hlim=64 time=0.417ms
received from 2001:db8:1:2::1, icmp_seq=4 hlim=64 time=0.415ms
received from 2001:db8:1:2::1, icmp_seq=5 hlim=64 time=0.417ms

6 packets transmitted, 6 packets received, 0.0% packet loss
round-trip min/avg/max = 0.414/0.421/0.449 ms
RTX1200#
RTX1200#show ipsec sa
Total: isakmp:2 send:2 recv:2


sa   sgw  isakmp  connection   dir    life[s]  remote-id
--------------------------------------------------------------------------------
1    1    -       isakmp       -      27937    2001:db8:0:2::2
2    1    -       tun[001]esp  send   27903    2001:db8:0:2::2
3    1    -       tun[001]esp  recv   27903    2001:db8:0:2::2
4    2    -       isakmp       -      28575    2001:db8:1:1::2
5    2    4       tun[002]esp  send   28703    2001:db8:1:1::2
6    2    4       tun[002]esp  recv   28703    2001:db8:1:1::2
RTX1200#
RTX1200#show ipv6 route
Destination             Gateway             Interface    Type
2001:db8:0:1::/64       -                   LAN1         implicit
2001:db8:0:2::/64       -                   LAN2         implicit
2001:db8:0:3::/64       -                   TUNNEL[1]    static
2001:db8:1:1::/64       -                   LAN3         implicit
2001:db8:1:2::/64       -                   TUNNEL[2]    static
```

■実習終了時設定内容

■ Cisco1841：

```
Current configuration : 1584 bytes
!
version 12.4
service timestamps debug datetime msec
service timestamps log datetime msec
no service password-encryption
!
hostname Cisco1841
!
boot-start-marker
boot system flash c1841-adventerprisek9-mz.124-24.T1.bin
boot-end-marker
!
logging message-counter syslog
!
no aaa new-model
dot11 syslog
ip source-route
!
ip cef
ipv6 unicast-routing
ipv6 cef
!
multilink bundle-name authenticated
!
archive
 log config
  hidekeys
!
crypto isakmp policy 10
 encr aes
 authentication pre-share
 group 2
crypto isakmp key 6 KEY1 address ipv6 2001:DB8:0:2::1/64
!
crypto ipsec transform-set test1 esp-aes esp-sha-hmac
!
crypto ipsec profile yamaha
 set transform-set test1
!
interface Tunnel0
 no ip address
 ipv6 unnumbered FastEthernet0/1
 tunnel source FastEthernet0/1
 tunnel destination 2001:DB8:0:2::1
 tunnel mode ipsec ipv6
 tunnel protection ipsec profile yamaha
!
interface FastEthernet0/0
 no ip address
 duplex auto
 speed auto
 ipv6 address 2001:DB8:0:3::1/64
 ipv6 enable
 no keepalive
!
```

Chapter 11 IPv6セキュリティ

```
interface FastEthernet0/1
 no ip address
 duplex auto
 speed auto
 ipv6 address 2001:DB8:0:2::2/64
 ipv6 enable
!
interface Serial0/0/0
 no ip address
 shutdown
 clock rate 125000
!
interface Serial0/0/1
 no ip address
 shutdown
 clock rate 125000
!
ip forward-protocol nd
no ip http server
no ip http secure-server
!
ipv6 route 2001:DB8:0:1::/64 Tunnel0
ipv6 route 2001:DB8:0:1::/64 2001:DB8:0:2::1
ipv6 route 2001:DB8:1:1::/64 2001:DB8:0:2::1
!
control-plane
!
line con 0
line aux 0
line vty 0 4
 login
!
scheduler allocate 20000 1000
end
```

■RTX1200：

```
# RTX1200 Rev.10.01.29 (Wed Feb  9 17:21:33 2011)
# MAC Address : 00:a0:de:66:ac:55, 00:a0:de:66:ac:56, 00:a0:de:66:ac:57
# Memory 128Mbytes, 3LAN, 1BRI
# main:   RTX1200 ver=b0 serial=D26036382 MAC-Address=00:a0:de:66:ac:55
  MAC-Address=00:a0:de:66:ac:56 MAC-Address=00:a0:de:66:ac:57
# Reporting Date: Jun 21 16:44:39 2011
console prompt RTX1200
ipv6 route 2001:db8:1:2::/64 gateway tunnel 2
ipv6 route 2001:db8:0:3::/64 gateway tunnel 1
ipv6 lan1 address 2001:db8:0:1::1/64
ipv6 lan2 address 2001:db8:0:2::1/64
ipv6 lan3 address 2001:db8:1:1::1/64
tunnel select 1
 tunnel encapsulation ipsec
 ipsec tunnel 100
   ipsec sa policy 100 1 esp aes-cbc sha-hmac
   ipsec ike encryption 1 aes-cbc
   ipsec ike group 1 modp1024
   ipsec ike local address 1 2001:db8:0:2::1
   ipsec ike local id 1 ::/0
   ipsec ike pre-shared-key 1 text KEY1
   ipsec ike remote address 1 2001:db8:0:2::2
ipsec ike remote id 1 ::/0
```

```
  tunnel enable 1
 tunnel select 2
  tunnel encapsulation ipsec
  ipsec tunnel 200
   ipsec sa policy 200 2 esp aes-cbc sha-hmac
   ipsec ike encryption 2 aes-cbc
   ipsec ike group 2 modp1024
   ipsec ike local address 2 2001:db8:1:1::1
   ipsec ike local id 2 ::/0
   ipsec ike pre-shared-key 2 text KEY2
   ipsec ike remote address 2 2001:db8:1:1::2
   ipsec ike remote id 2 ::/0
  tunnel enable 2
 ipsec auto refresh on
RTX1200#
```

■ AX620R：

```
Using 1561 out of 524288 bytes

! NEC Portable Internetwork Core Operating System Software
! IX Series IX2025 (magellan-sec) Software, Version 8.5.21, RELEASE SOFTWARE
! Compiled Sep 03-Fri-2010 10:15:21 JST #1
! Last updated Jun 21-Tue-2011 16:35:58 JST
!
hostname AX620R
timezone +09 00
!
ipv6 route 2001:db8:0:2::/64 2001:db8:1:1::1
ipv6 access-list list1 permit ip src any dest any
!
ike proposal ike-prop encryption aes hash sha group 1024-bit
!
ike policy ike-policy peer 2001:db8:1:1::1 key KEY2 ike-prop
!
ipsec autokey-proposal ipsec-group esp-aes esp-sha
!
ipsec autokey-map ipsec-policy list1 peer 2001:db8:1:1::1 ipsec-group
ipsec local-id ipsec-policy 2001:db8:1:2::/64
ipsec remote-id ipsec-policy 2001:db8:0:1::/64
!
device FastEthernet0/0
!
device FastEthernet0/1
!
device FastEthernet1/0
!
device BRI1/0
  isdn switch-type hsd128k
!
interface FastEthernet0/0.0
  no ip address
  ipv6 enable
  ipv6 address 2001:db8:1:2::1/64
  no shutdown
!
interface FastEthernet0/1.0
  no ip address
  ipv6 enable
  ipv6 address 2001:db8:1:1::2/64
```

```
   no shutdown
!
interface FastEthernet1/0.0
  no ip address
  shutdown
!
interface BRI1/0.0
  encapsulation ppp
  no auto-connect
  no ip address
  shutdown
!
interface Loopback0.0
  no ip address
!
interface Null0.0
  no ip address
!
interface Tunnel0.0
  tunnel mode ipsec
  no ip address
  ipv6 enable
  ipv6 unnumbered FastEthernet0/0.0
  ipsec policy tunnel ipsec-policy out
  no shutdown
AX620R(config)#
```

索 引

数字・アルファベット

6to4 (RFC3056) ･･････････････････････ 10
ACL ･････････････････････････････････ 168
DHCP ･････････････････････････････････ 3
DHCPv6 サービス ･･･････････････････ 24
DNS ･･･････････････････････････････････ 9
IPsec ･････････････････････････････ 9, 172
IPv6 over IPv4 トンネリング ･･････ 150
IPv6 逆引きファイル ･･･････････････ 162
IPv6 正引きファイル ･･･････････････ 161
IPv6 ローカル逆引きファイル ･････ 162
IPv6 ローカル正引きファイル ･････ 162
ISATAP (RFC5214) ･･････････････････ 10
O フラグ ･･･････････････････････ 24, 34
OSPFv3 ･････････････････････ 8, 92, 106
 ── データベース ･･･････････ 100, 126
RA (Router Advertisement) ･･････ 3, 34
 ── の設定 ･･････････････････････ 23
RFC5952 ･･････････････････････････････ 3
RIPng ･･････････････････････････ 7, 62, 74
 ── プロセスの起動 ･･････････ 63, 69
 ── メッセージ送信の制御 ･･･ 64, 69
Teredo (RFC4380) ･･････････････････ 10
Web ･･･････････････････････････････････ 9

あ行

アクセスレベル ･･････････････････ 16, 17
一般ユーザー ･･･････････････････････ 16
イネーブルモード ･････････････ 15, 16, 18
エニーキャスト ･･･････････････････････ 3
エニーキャストアドレス ･･････････････ 4
オペレーションモード ･･･････････････ 16

か行

管理ユーザー ･･･････････････････ 16, 18
グローバルコンフィグレーションモード ･･･ 15
グローバルユニキャストアドレス ･･････ 4
コンフィグモード ･････････････ 16, 17, 18

さ行

サイトローカル ･･･････････････････････ 3
スタティックルート ･････････････････ 42
ステートフル構成フラグ ･･･････ 24, 34
ステートフル割り当て ･･･････････････ 6
ステートレス割り当て ･･･････････････ 6

た行

デフォルトルート ･･･････････････････ 52
 ── の伝搬 ･････････････････････ 118

な行

ネットワークアドレス ･･･････････････ 3
ネットワークトポロジー ･･･････････ 42
ノードアドレス ･･･････････････････････ 3

は行

プレフィックス ･･････････････････ 3, 23

ま行

マルチキャスト ･･･････････････････････ 3
マルチキャストアドレス ･･･････････ 4, 5
マルチキャストグループ ･････････････ 5
メールサーバー ･････････････････････ 164

や行

ユーザーモード ･･････････････････････ 15
ユニキャスト ･････････････････････････ 3
ユニークローカルユニキャストアドレス ･･･ 4

ら行

リンクローカル ･･･････････････････････ 3
リンクローカルユニキャストアドレス ･･･ 4
隣接情報 ･･････････････････････ 99, 125
ルーター ID ･･･････････ 93, 106, 118, 136
ルーターの初期化 ･･････････････････ 20
ルーティングテーブル ･････････ 42, 62

著者紹介

前野 譲二(まえの じょうじ)
　早稲田大学 メディアネットワークセンター 助教
　特定非営利活動法人 インターネット・ラーニングアカデミー 理事(～2012年8月)

鈴田 伊知郎(すずた いちろう)
　アラクサラネットワークス株式会社 営業本部 主任
　特定非営利活動法人 インターネット・ラーニングアカデミー
　次世代ネットワークの構築と教育プロジェクト リーダ(～2011年5月)

小林 貴之(こばやし たかゆき)
　日本大学文理学部 准教授
　特定非営利活動法人 インターネット・ラーニングアカデミー 理事

IPv6ネットワーク構築実習
Practical Studies : Building IPv6 Network

2013年 4月25日　初版1刷発行

著　者　前野 譲二・鈴田 伊知郎・小林 貴之　© 2013　　　　　（検印廃止）

発行所　**共立出版株式会社** ／南條光章

　　　　東京都文京区小日向4丁目6番19号
　　　　電話　03-3947-2511番（代表）
　　　　〒112-8700／振替口座 00110-2-57035番
　　　　URL　http://www.kyoritsu-pub.co.jp/

一般社団法人 自然科学書協会 会員
NDC 547.48
ISBN 978-4-320-12335-9
Printed in Japan

印刷・製本：錦明印刷　　本文組版・装丁：IWAI Design

JCOPY ＜(社)出版者著作権管理機構委託出版物＞
本書の無断複写は著作権法上での例外を除き禁じられています。複写される場合は、そのつど事前に，(社)出版者著作権管理機構（電話 03-3513-6969, FAX 03-3513-6979, e-mail: info@jcopy.or.jp）の許諾を得てください。